A Lifetime of Puzzles

A Collection of Puzzles in Honor of Martin Gardner's 90th Birthday

Edited by
Erik D. Demaine
Martin L. Demaine
Tom Rodgers

A K Peters, Ltd.
Wellesley, Massachusetts

Editorial, Sales, and Customer Service Office

A K Peters, Ltd.
888 Worcester Street, Suite 230
Wellesley, MA 02482
www.akpeters.com

Library of Congress Cataloging-in-Publication Data

A lifetime of puzzles / edited by Erik D. Demaine, Martin L. Demaine, Tom Rodgers.
    p. cm.
  Includes bibliographical references and index.
  ISBN-13: 978-1-56881-245-8 (alk. paper)
  1. Mathematical recreations. 2. Gardner, Martin, 1914- I. Demaine, Erik D., 1981- II. Demaine, Martin L., 1942- III. Rodgers, Tom, 1943-
QA95.L73 2008
793.74–dc22

                                        2007041016

Printed in India
12 11 10 09 08   10 9 8 7 6 5 4 3 2 1

Contents

Preface

This book celebrates Martin Gardner's 90th birthday with a series of 25 articles about some of Martin's favorite topics.

Martin Gardner is the father of recreational mathematics, an avid puzzler, a lifelong magician, and a debunker of pseudoscience. He has written more than 65 books throughout science, mathematics, philosophy, literature, and conjuring. He has deeply influenced countless readers of his "Mathematical Games" column in *Scientific American*, which ran for 25 years from 1957 to 1982. This column popularized recreational mathematics and introduced many connections between mathematics, puzzles, and magic. Together with Gardner's amazing ability to correspond with his many readers, the columns gave the general public the opportunity to enjoy mathematics and to participate in mathematical research. Many of today's mathematicians entered this field through Gardner's influence. A whole body of research into recreational mathematics has also emerged, solving problems that Gardner posed years ago and introducing new problems in the same spirit.

Given the retrospective nature of this book, many of the articles have a historical slant. The first two articles, for example, are specifically about Martin Gardner and his influence on the world of magic. Part II is entirely "In Hindsight," describing the world's first puzzle "craze" of the *Tangram* and detailing the oldest book on recreational mathematics (circa 1500) including both puzzles and magic tricks. Several articles consider historical puzzles; for example, Roger Penrose reminds us of a kind of maze he developed with his son back in 1958. (Incidentally, the present book was also edited by a father-son team.)

The articles in this book are organized into six parts. Part I, "Cast a Spell," is about mathematical magic tricks. Part II, "In Hindsight," makes the historical discoveries described above. Part III, "Move It," is about puzzles involving motion, from mazes to *Instant Insanity* to Peter Winkler's walking ants. Part IV, "Fitting In," is about puzzles involving packing or entanglement, from Stewart Coffin's work to burr puzzles, and the related art of mosaics. Part V, "Speak to Me," is about word puzzles, from Smullyan's logic puzzles to recreational linguistics on graphs and grids. Part VI, "Making Arrangements," is about puzzles and games that arrange pieces into particular structures, from the classic Gardner topics of ticktacktoe and magic squares to new developments inspired by Gardner (or Dr. Matrix) like pandigital numbers.

We feel honored to gather this collection of exciting and fun material in honor of a man who has touched so many: Martin Gardner.

Erik D. Demaine
Cambridge, Massachusetts

Martin L. Demaine
Cambridge, Massachusetts

Tom Rodgers
Atlanta, Georgia

Part I

Cast a Spell

Warning: Martin Gardner has turned hundreds of mathematicians into magicians and hundreds of magicians into mathematicians! ~Persi Diaconis

Martin Gardner and His Influence on Magic

Christopher Morgan

Persi Diaconis thinks the good-natured "warning" noted above should appear on many of Martin Gardner's books. A distinguished Stanford mathematician, magician, and long-time friend of Martin Gardner, Persi has written several technical treatises on the mathematics of card shuffling, among his many other accomplishments. He has tremendous admiration for Martin Gardner: "Martin elevates magic in our eyes and in the public's eyes. He's such a visible center that people from all over the world have written to him. He picks the best ideas and amplifies them."

Many magicians who know and love Martin Gardner's magical writings would agree. This short essay discusses Martin's accomplishments in magic and their connections to mathematics, mostly through the voices of those many "mathemagicians" who have been influenced by him.

A Focal Point

Martin Gardner stands at the intersection between magic and mathematics. "Mathematical magic, like chess, has its own curious charms," he says. "[It] combines the beauty of mathematical structure with the entertainment value of a trick" [7]. Persi Diaconis understands this synthesis:

> The way I do magic is very similar to mathematics. Inventing a magic trick and inventing a theorem are very, very similar activities in the following sense. In both subjects you have a problem you're trying to solve with constraints. One difference between magic and mathematics is the competition. The competition in mathematics is a lot stiffer than in magic. [3]

Many of Martin Gardner's fans may not know the extent of his lifelong involvement in magic, or how many contributions he has made to the art. Indeed, many future magicians began reading Martin Gardner for the mathematics, only later becoming fascinated by the magical content. Magician Dan Garrett [8], for example, grew up with an interest in mathematics and science rather than magic (other than as a hobby). In high school, he says, "I read Martin's 'Mathematical Games' column in *Scientific American* and his book *The Numerology of Dr. Matrix*. I never even knew he was a magician until much later." (He notes that Gardner's

Figure 1. Martin Gardner showing Joe Berg's improved version of the Hunter rope trick.

Encyclopedia of Impromptu Magic [5] is a tremendously significant contribution to the vast world of magic literature.)

Martin has always been generous with his magical ideas. Colm Mulcahy, a professor in the Department of Mathematics at Spelman College and creator of an excellent website devoted to mathematical card tricks [11], told me that over a period he gradually became fascinated by mathematical card tricks and ultimately started corresponding with Martin Gardner, who graciously allowed him to recycle any of the card tricks in his mathematics popularization publications and even suggested that Colm write a book on the subject of mathematical card tricks, which he is now doing.

Magic has always been Martin's main hobby, and he pursues it actively to this day. In *Martin Gardner Presents* [6, p. 374], a comprehensive 1993 collection of Martin's magical creations, Stephen Minch (magician, author, and founder of Hermetic Press) notes that "card magic, and magic in general, owe a far greater debt to Martin Gardner than most conjurors realize." Martin was recognized for these contributions in 1999, when he was named one of *MAGIC Magazine's* 100 most influential magicians of the twentieth century [1].

Seven Decades of Magic

Martin has been writing about magic and contributing new effects for nearly seventy years. His magical friends past and present have included Dai Vernon ("The man who fooled Houdini"), Persi Diaconis, Jerry Andrus, Stewart James, Wesley James, Ed Marlo, Doctor Daley, Mel Stover, Ted Annemann, Ken Krenzel, Max Maven, Howie Schwarzman, Jay Marshall, Richard Kaufman, Herb Zarrow, Karl Fulves, and many, many others. Now, in his nineties, he keeps in contact with magicians like Penn and Teller by phone and receives occasional visits from magicians who come to trade notes with him.

I visited Martin recently to discuss his career in magic—which we did, though we actually spent more time trading magic tricks! His enthusiasm for new magical ideas remains as infectious as ever.

The spriest of nonagenarians, Martin showed me Joe Berg's improved Hunter knot trick [6, p. 36], two false deck cuts, a revolving card effect, some topological knot tricks, some rubber band tricks, and several mathematical card tricks. Many of these tricks have appeared in his writings over the years. Next, he demonstrated the Wink Change, an elegant card effect he created years ago. "Of all the moves I have invented," he said, "the Wink Change is the one

Figure 2. Martin Gardner demonstrating his original card effect, the Wink Change.

I'm most proud of" [6, p. 315]. The Wink Change instantly transforms one card into another, and he performed it with the kind of effortless technique that comes only from years of practice. Before we knew it, the afternoon had flown by.

Martin prefers the intimate approach to magic:

> I'm not a performer. I just do close up stuff for friends. The only time I got paid for doing any magic was when I was a student at the University of Chicago. I used to sell magic sets at Marshall Fields during the Christmas season. One of the Gilbert magic sets had some pretty nice apparatus in it, and I worked out several routines. That's the only time I had to do magic in front of a crowd.

Early Work and a Meeting with Annemann

Martin's first published magic manuscript, *Match-ic*, appeared in 1935. It was a booklet of tricks featuring matches. Many more pamphlets would follow. Several of them, including *12 Tricks with* a *Borrowed Deck* and *Cut the Cards,* are highly regarded by magicians.

One standout trick from *12 Tricks with a Borrowed Deck* is Martin's "Lie Speller" trick [4, 6, p. 172]. In this effect, a spectator secretly picks a card, replaces it in the deck, and ultimately spells the name of the card, one card per letter. Amazingly, the last card dealt is the spectator's card. This works even if the spectator lies about the name and/or suit of the card! Variations on this seminal trick have been created by such magicians as Jack Avis, Bruce Cervon, Milt Tropp, Harvey Rosenthal, Larry Jennings, Ed Marlo, Jon Racherbaumer, Max Maven (a.k.a. Phil Goldstein), Bob Somerfeld, Allan Slaight, Stewart James, and J. C. Wagner, among others.

A turning point for Martin was his first meeting with Ted Annemann in the late 1930s. Annemann, editor of the influential magic magazine *The Jinx,* was one of the most fertile minds in magic and mentalism during the first half of the twentieth century. Magician Steve Beam, author of the excellent *Semiautomatic Card Tricks* book series, notes that Annemann was an important early pioneer in mathematically based magic tricks. "Annemann hid many mathematical principles in his card tricks," he says. Many of these mathematical tricks appeared in *The Jinx* during the 1930s and 1940s, and Martin's original effects were among them.

Martin recalls his first meeting with Annemann in New York in 1937:

> I had just recently moved from Chicago to New York. I walked into a bar restaurant on Broadway and 42nd Street and recognized Annemann sitting at a table with Doc Daley. I recognized him from his photo. I had just published a book of card tricks, and I told Annemann I had a manuscript for a new book of original ideas I had in magic. He invited me to come to his apartment and demonstrate some of the tricks. He had a little stage at one end of his apartment [laughs]. I stood up there and did a series of tricks. He asked if he could borrow the manuscript, and he devoted an issue to the tricks that were in that manuscript. So, my Lie Speller trick first appeared in *The Jinx.*

Annemann historian Max Abrams says:

> *The Jinx 1937–1938 Winter Extra* consisted of 24-year-old Martin Gardner's "Manuscript," an eight-page bonanza containing seventeen tricks by the prolific and profound Martin Gardner. The collection of tricks presented in the *Extra* was an auspicious occasion in a redoubtable career. [2, p. 364]

Mathematics, Magic and Mystery

Ask Gardner magical aficionados to name their favorite Gardner book, and you'll often hear *Mathematics, Magic and Mystery* [7]. Martin's first published book, it appeared in 1956 and remains in print a half century later. This seminal work has been a bible for magicians interested in mathematically based tricks.

Max Maven, one of the most creative mentalists and magicians in the field, wrote the introduction to *Martin Gardner Presents*. He keeps two copies of *Mathematics, Magic and Mystery* in his library. That's because over the years he has referred to his first copy so often that it has begun to fall apart. (He holds on to that first copy "for sentimental reasons." Thus far, the later reprint is holding up to frequent handling.) He told me:

> Martin is one of the great teachers, not only of magic, but of science and mathematics. Although Martin's work in magic is not primarily invention, he has in fact devised some excellent material, and several of his creations (most notably the Lie Speller, both for plot and method) have become standards. But his great gift is gathering really good information, separating the wheat from the chaff, then explaining those ideas with writing skills that make them engaging and understandable. He has been a conduit—perhaps a better word is "synthesizer"—for a great deal of magical information that has filtered out into the larger magic world.
>
> A lot of his influence has been secondary and tertiary, simply because many people who've come up in magic more recently have not realized that they were being influenced by him. That's because a lot of his ideas, or the ideas he was conveying, had already passed through other people. In my case, I had the benefit of having a father who was a physicist, and therefore I was reading Martin's [*Scientific American* "Mathematical Games"] column from a very early age. I owe a lot to Martin Gardner, for expanding my intellectual horizons.

Mathematics, Magic and Mystery is particularly valued because it records some of the best mathematical tricks of the eccentric magical genius Bob Hummer, whose parity-based card tricks inspire magicians to create interesting variations to this day. The Hummer effects are just a few of the riches to be found there. Martin's elegant "Curry Triangle" (a deceptive geometrical vanish effect) also appears there, for example [7, p. 145].

Patterns and Principles

Ken Krenzel, another respected name in the field of magic, is an old friend of Martin's. Ken told me:

> Martin is brilliant. He has always been one of my magical heroes. He's a very quiet, almost shy person. When I first met him in New York [in 1956], I saw him in the New York Public Library on 42nd Street. He was always up in the reference room, collecting material. That was back when he published his first book, *Mathematics, Magic and Mystery*. The depth and breadth of Martin's writing is just incredible. If you look in Hugard's *Encyclopedia of Card Tricks* [9, p. 167], for example, you'll see a magnificent, little-known subtlety under his name, called "Gardner's Unique Principle." It involves one-way backs. His subtlety is that you can have the cards mixed every which way, with the backs facing randomly in both directions. When you spread the cards face down in a ribbon spread on the table, you look for patterns, such as four cards going in one direction followed by three going in the other, then perhaps five in the other, and so on. That gives you your key as to where a card is taken or replaced.

Magical Secrets and "Elevating Magic"

The aforementioned Steve Beam says that he's always been a big fan of Martin Gardner, and that reading him also got him hooked on math. Over the years, Steve has used a highlighter on so many passages in books like Gardner's *Mathematical Magic Show* that the pages are completely yellow. For magicians, he notes, one of the most attractive aspects of Martin's writing is the emphasis on elegant principles rather than finished effects. Martin encourages the readers to embellish the ideas: "There's a lot of great raw material in Gardner's writings," he notes. "People can run with his material because he doesn't work it to death." This emphasis on theory may be one reason, says Persi Diaconis, that magicians are seldom bothered when Martin reveals elegant "mathemagical" ideas in his columns and books:

> In magic, secrets are sacrosanct, yet Martin has routinely put wonderful secrets into his columns, and somehow the world forgives him. Part of the reason is that having magic presented in his *Scientific American* column or in his books—in the glow of so many other important ideas—glorifies magic. People are proud to have a trick in one of Martin's books. In fact, my first

published magic trick appeared in Martin's column years ago. And, because he is able to elicit magic from unlikely sources, often outside of the magic world, many unexpected new mathemagical ideas have come to light.

A good example of this is the Kruskal principle [10], invented by Princeton physics professor Martin Kruskal, and used in many card effects. It is based on—of all things—Markov chains!

Colm Mulcahy adds, "Other ideas with far-reaching consequences that Martin has introduced to the reading world at large are faro (perfect) card shuffles and the Gilbreath principle." The Gilbreath principle is named after its inventor, Norman Gilbreath. In its simplest form, if you arrange a deck of cards so that the colors alternate, cut the deck into two halves so that the bottom cards are different colors, then riffle shuffle the halves together once, each pair of cards will contain one red card and one black card. Colm Mulcahy tells me that early in the twentieth century, O. C. Williams published the basic fact that a single irregular riffle shuffle falls far short of randomizing a deck of cards. This was later expanded on by Charles Jordan. In the late 1950s, Norman Gilbreath and others rediscovered the principle and took it to new heights. Karl Fulves also says that Gene Finnell independently discovered the principle.

Dignity for Our Little Mysteries

Gordon Bean, well-known magician and magical author, says:

> As the son of a physicist, I lived in a house visited regularly by Martin Gardner's column in *Scientific American*. After my interest in magic kindled, I can remember few satisfactions greater than the tantalizingly infrequent times "Mathematical Recreations" would veer into the realm of magic. Apart from the actual principles and effects explored, I've never escaped the reverberations of visiting a place where being able to magically predict the position of red and black cards after a legitimate shuffle seemed only a little less important than being able to predict the rotation of planets.
>
> English-speaking magicians have long been frustrated by the inadequacy of the word "trick" to describe what they do. This is a lack that we'll most likely never fill, but Martin Gardner has gone a long way in bringing our little mysteries a sense of dignity without ever losing a sense of fun.

Figure 3. A portrait of Martin Gardner.

A Clarity of Perception

Atlanta-based magician Joe M. Turner [12] offers a fitting conclusion to this essay by putting things in a larger context:

> Martin Gardner's long and continuing influence in magic is—if you'll pardon the pun—puzzling. After all, some of the most common advice we magicians give ourselves is to "perform magic" and not just to "do tricks." We are encouraged to enchant and mystify our audiences by creating a theatrical experience, and to lift them above the perception of magic as a "mere puzzle."
>
> And yet, Martin Gardner remains one of the most cited and revered names in our field. Martin Gardner! His *Encyclopedia of Impromptu Magic* is guaranteed to show up in any poll of magicians' favorite magic books. His magazine columns are the source of endless fascination among magicians as well as actual human beings. Throughout his work we find items which bear frighteningly close resemblance to (gasp!) puzzles. Why does a mathematician with a predilection for impromptu tricks and puzzles command so much attention that magicians jockey to get invited to a convention named in his honor? It must be more than simply the prestige of telling other magicians you were there.
>
> Perhaps Martin Gardner, for all the perception-twisting puzzles and tricks he has created, has a clarity of perception with regard to magic that transcends even what magicians understand our art to be. Magic is more than the special effects we see on a stage or in the practiced hands of a trickster, however talented. Martin Gardner shows us that magic, like mathematics, may in fact be an intrinsic and often surprising part of how the universe is put together. Just when we think we've got something figured out, he shows up with a different way of looking at it and we are surprised by the very thing we thought we knew—whether it's a mathematical principle, a deck of cards, or a piece of string. Martin Gardner reveals the surprising in the familiar, which—if one wishes to create the illusion of magical powers—is a skill devoutly to be wished.

Acknowledgments. Thanks to the many magicians and mathematicians who have graciously helped with this tribute. They include Persi Diaconis, Max Maven, Steve Beam, Arthur Benjamin, Joe M. Turner, Stan Allen, Colm Mulcahy, Howie Schwarzman, Ken Krenzel, Gordon Bean, and Dan Garrett. And, of course, I thank Martin Gardner, for allowing me into his home and showing me such great magic!

Bibliography

[1] "100 Most Influential Magicians of the 20th Century." *MAGIC Magazine*, June 1999.

[2] Max Abrams. *The Life and Times of A Legend: Annemann*. Tahoma, CA: L&L Publishing, 1992.

[3] BCC faculty. "Persi Diaconis." *Mathographies*. Available at http://scidiv.bcc.ctc.edu/Math/Diaconis.html. Condensed biography from *Mathematical People*, edited by Donald J. Albers and Gerald L. Alexanderson. Boston: Birkhauser, 1985.

[4] Joe Berg. *Here's New Magic: An Array of New and Original Magical Secrets*. Chicago: Joe Berg, 1937.

[5] Martin Gardner. *Encyclopedia of Impromptu Magic*. Chicago: Magic, Inc., 1978.

[6] Martin Gardner. *Martin Gardner Presents*. New York: Kaufman and Greenberg, 1993.

[7] Martin Gardner. *Mathematics, Magic and Mystery*. New York: Dover Publications, 1956.

[8] Dan Garrett's homepage. Available at http://members.aol.com/dangarrett/, accessed 2007.

[9] Jean Hugard, editor. *Encyclopedia of Card Tricks*. New York: Dover Publications, 1974. (Ken Krenzel notes that this book was mostly the work of Glenn G. Gravatt, a.k.a. Doctor Wilhelm von Deusen.)

[10] Jeffrey C. Lagarias, Eric Rains, and Robert J. Vanderbei. "The Kruskal Count." Available at http://arXiv.org/abs/math/0110143, 2001.

[11] Colm Mulcahy. "Colm's Cards—Mathematical Card Tricks." Available at http://www.spelman.edu/~colm/cards.html, 2007. See also Colm Mulcahy, "Low Down Triple Dealing," *Focus: The Newsletter of the Mathematical Association of America*, 24:8 (November 2004), 8, available at http://www.maa.org/features/tripledeal.html.

[12] Joe M. Turner. "Joe M. Turner: America's Corporate Magic Communicator." Available at http://www.joemturner.com, 2006.

Martin Gardner—Encore!

Prof. M. O'Snart

What is this eminent scholar and philosopher doing in the pages of *MAGIC*?[1] He is following his lifetime passion—sharing the mystery, the surprise, and the joy of Magic!

The Challenge Set

The most mysterious figure in the realm of magical literature, whose one contribution to the subject is still, after 25 years, one of the classics, is S. W. Erdnase, author of *The Expert at the Card Table*. Who was S. W. Erdnase? It has been said that his real name was E. S. Andrews, which in reverse order produces the pen name under which he wrote.

This challenge was laid down by Leo Rullman in *The Sphinx* for February 1929 and was not met in his lifetime. But the puzzling wordplay and the darker mystery of authorship would have struck one very inquisitive 14-year-old student of magic very deeply: Martin Gardner's subconscious stored the challenge away, biding its time.

Rullman was soliciting lists of "The Ten, or Twenty, Best Books on Magic" in his column, and Erdnase appeared in most; as a

---

[1]The original version of this article appeared in *MAGIC Magazine*, April 2004. Reprinted with permission. www.magicmagazine.com

book dealer, he listed a "scarce original edition" for $2.25; Dariel Fitzkee's new book, *Jumbo Card Manipulation*, was hailed as "The Erdnase of Jumbo Cards"—Erdnase was hot! Martin was well immersed in the jog shuffles and fancy cuts, the ruse and subterfuge of Erdnase. He read his 25¢ copy "with passionate interest." And the next year, he would be writing in *The Sphinx* himself!

Tulsa, Oklahoma

Martin was born October 21, 1914 in Tulsa, Oklahoma. His father, Dr. James H. Gardner, geologist and oilman, taught Martin his first trick, "Papers on the Knife Blade." Encouraged by Roy "Wabash" Hughes, Roger Montandon, and Logan Wait, he advanced rapidly in the art. Much later, Montandon and Wait would include two items from Martin in their 1942 booklet, *Not Primigenial*, commenting, "We've always enjoyed Martin Gardner's quick tricks." In 1978, Martin dedicated his massive *Encyclopedia of Impromptu Magic* "For Logan and Roger."

While 15, Martin contributed nine effects to *The Sphinx*: "New Color Divination" (of gum balls) in May 1930, to "The Travelling Stick of Gum" and "Vanishing Pack of Life Savers" in October 1930. Fellow contributors and dealers that year were Stewart James, with a new card effect, "Gnikool," for 50¢ and John Booth, introducing his original "Three Shell Monte," also for 50¢. Beneath Martin's second contribution, "Borrowed Ring Off String," Charles "Baffles" Brush in his "Current Magic" department for July 1930, commented, "This is the right way to do magic, take one trick and use if for an entirely different one. Give him a little encouragement and he will send in another good one." Did Baffles hit the old prophetic nailhead right in the bull's eye, or what? Since that first display of creativity in 1930, every single year has seen some published work of Martin's!

In the August 1930 issue of *The Sphinx*, Martin described his own subtle and effective changes in handling for the "Papers on the Knife." In 1978, he devoted six pages to this classic in his *Encyclopedia*. Martin's name appeared for the first time on the front cover of *The Sphinx* as a contributor (along with ten others) in the February 1931 issue. Of course, he was now 16!

The March 1931 issue of *The Sphinx* was a special issue, marking the magazine's 30th year with "the first rotogravure section ever found in a magic periodical" of eight pages of portraits. Martin contributed "An Impromptu Trick" in which a borrowed and marked coin travelled from pocket to opposite pant leg cuff twice. This title must have amused Martin, since it required adding a secret pocket and a two-foot-long cloth coin slide to your trousers! I suspect this was a put-on, since this elaborate method was used by Martin as the mock explanation for a similar effect done by simple sleight-of-hand in the 1937–38 *The Jinx Winter Extra*. This contribution also marked Martin's debut as a dealer. For ten cents postage, he offered, "a pair of black elastic shoelaces for use in seance work where it is necessary to remove foot from shoe. They cannot be told from the genuine article and save lots of time in taking shoe on and off." Baffles kindly let this blatant commercial message slip by—after all, those were Depression days!

Chicago, Illinois

Martin left Tulsa for Chicago in 1932, expecting to spend two years at the University of Chicago, then shift to Caltech and complete his education as a physicist. However, he found the philosophy of science and philosophy in general so attractive that he remained to graduate Phi Beta Kappa in philosophy from the University of Chicago in 1936. Fiction, poetry, philosophy, and politics engaged Martin's flowering writing talents, but chess engaged his spare time to an extent so alarming that he decided to quit playing completely rather than become compulsive.

In November 1935, the Ireland Magic Company of Chicago published Martin's first booklet, *Match-ic*, "More Than Seventy Impromptu Tricks With Matches."

Martin returned to Tulsa in 1936 for a stint as a reporter for the *Tulsa Tribune*, did not like it, and returned quickly to Chicago and the action at Joe Berg's and Laurie Ireland's magic shops. His day job was public relations writing for the University of Chicago, but Martin fondly recalls that he was "a charter member of the old Round Table gang that used to meet every night at the Nankin Chinese restaurant on Randolph Street. Werner 'Dorny' Dornfield was the group's central figure, and I count my friendship with him as one of the great privileges of my youth."

Ed Marlo, or Eddie "Bottom Deal" Marlo back then, was another good friend of Martin's from the 1930s. Their joint effort, the "Gardner-Marlo Poker Routine" in Marlo's 1942 booklet *Let's See The Deck*, became the classic automated model. Another classic card plot created by Martin, "The Lie Card Speller," wherein the spectator may lie or tell the truth to every question, first appeared in *Here's New Magic*, published in 1937 by Joe Berg, ghosted by Martin Gardner.

L. L. Ireland published Martin's *12 Tricks with a Borrowed Deck* in 1940, full of remarkably durable material. Martin published (that is, mimeographed) his own manuscript, *After The Dessert*, in 1940. Its quick success justified a printed edition, expanded from 24 to 30 impromptu tricks, which Max Holden published in 1942, dedicated to "Dorny." The title page is graced by a quote from William Shakespeare: "After the Dessert ... 'Tis a Goodly Time for Pleasantry." Martin finally admitted it was just another of his spoofs—he made up the quotation after he made up the title! The *Genii* ad of December 1940 has the name of the author/dealer as "Matt Gardner," possibly to distinguish orders generated from the ad in *The Sphinx* of November 1940?

In the late 1930s, Martin became known as an idea man, always ready to generate fresh material for novelty houses, stories or articles for publishers, and ideas for cereal box inserts or merchandising premiums. For 1938, 1939, and 1940, Martin worked at Marshall Field's department store, demonstrating and selling "Mysto Magic Sets" through the Christmas season. Martin says he learned there that you don't know a trick until you have performed it 50 times. Martin invented a transposition effect using two large sponge balls in 1940, which Ireland marketed as "Gardner's Passe Passe Sponge Trick," four pages plus two sponges for 50. Martin must know this effect extremely well. At the Chicago 1940 SAM Convention, he performed the moves "a few thousand times," pitching the package in the dealers' room. No report as to the number of sales made.

The North Atlantic

Martin enlisted in the Navy in 1941 and served four years on a destroyer escort, the USS Pope, with the North Atlantic Fleet. He spent much of his night-watch time thinking up plots for stories, much like Stewart James in the Canadian Army, who volunteered for night-guard duty so that he could work out his magical methods without interruption. The year 1942 saw Max Holden publish Martin's second booklet on cards, *Cut the Cards*. In the introduction, Martin worried that "tricks will continue to be forgotten and later reinvented, or to be buried permanently in some remote corner of an old magazine or out-of-print booklet." It is ironic that Martin's other booklets were all reprinted many times, but *Cut the*

Cards went out of print for 50 years. It may now be found, eccentrically displayed, in *Martin Gardner Presents*, 1993. To solve the problem, Martin envisaged "a mammoth card encyclopedia" wherein "thousands of sleights, principles, and effects will be described, classified and cross-indexed . . . Additions to the book will appear annually as pamphlets, and at intervals, the entire work will be revised and reissued."

The Challenge Met

In 1945, Martin returned to civilian life, freelance writer style, in Chicago. The next year, provoked perhaps by news of the death of Leo Rullman or by some chance remark, an urge grew to take up that long-ignored challenge, "Who was S. W. Erdnase?" Fortune smiled on Martin and his oily old pea-jacket! In December 1946, Martin found and met Marshall D. Smith, the actual illustrator of *The Expert at the Card Table*, who had drawn the very hands of Erdnase demonstrating his sleights! An elated Martin arranged a guest appearance of M. Smith at the 1947 SAM Convention in Chicago, where he met Erdnase enthusiasts and autographed their copies. Alas! None of the leads to Erdnase so hoped for came from the artist.

But Martin soon had a new lead. In the August 1949 *Conjuror's Magazine*, he proffered new evidence—an article by James Andrews from the June 26, 1909 *Harper's Weekly* entitled "The Confessions of a Fakir." Martin wrote that he could not prove it, but he thought Andrews was Erdnase. But again, no real ties were found.

Fortunately, Walter Gibson supplied Martin with a lead to an old gambler, Ed Pratt, who had known Erdnase! His recollections provided the needed clues for the identification of Erdnase as Milton F. Andrews, achieved October 29, 1949! By November, Martin had found and interviewed Milton's older brother, Alvin E. Andrews. In *The Phoenix* #190, November 18, 1949, Bruce Elliott announced that Martin had solved the case:

> A really exciting thing has happened—Martin Gardner's lone, long quest for the truth about the mysterious author of *The Expert at the Card Table* has been crowned by success. Photostats being airmailed to New York will show once and for all who the strange Mr. Erdnase really was, what his life was like, and how and when he died.
>
> We're hoping to be able to bring you the highlights of the story in the next issue. If we were able to tell you the story, you'll

agree that it is one of the most remarkable ever told. Even if we can't, we must doff our dusty lid to Martin and his stick-to-itiveness against seemingly insurmountable odds. Many have tried to find out about Erdnase. Only Martin, working from vague hints and even vaguer clues and hunches, has seen his way through the web of misinformation.

Pictures, news stories, a confession all iron clad evidence provide that Martin has solved the case. When we saw Martin, he had just come from interviewing a blood relative of Mr. Erdnase. In the excitement of the chase, Martin hadn't had a chance to change his clothes for three days. Even a literary detective case can have its wild moments. This was the culmination for Martin of years of work, of probing the libraries, checking city directories, of adding two and two and getting fourteen.

We envy him.

In the next issue Elliott could only say, "Still haven't received clearance on the Erdnase story. Maybe in next issue." Five years would go by before another word on the strange Mr. Erdnase saw print. Martin preferred to spend more time gathering data and verifying details with Pratt, Smith, and others. Even the cemetery where Erdnase lay buried was checked. Finally, on December 24, 1954, Jay Marshall in *The New Phoenix* #321 announced the true identity of Erdnase:

Martin Gardner brought with him a briefcase and a sheaf of photostats. It was a complete newspaper account of the exciting life and the dramatic suicide of Milton Franklin Andrews. He had the correspondence and the notes made during the last decade in his successful search for the true identity of the elusive idol of the card sharps: S. W. Erdnase. It's a story of crime, murder, and adventure that is stranger than fiction. You'll find it all in these pages during the coming Summer.

In *The New Phoenix* #339, September 1956, Jay Marshall noted, "We are still at work on the Erdnase story and hope to publish it in full sometime this fall." By the following issue #340, January 10, 1957, plans had changed. "Martin Gardner is going to write the Erdnase story for *True* magazine."

When *True* published Martin's account of Erdnase in January, 1958, entitled "The Murdering Cardshark," it had been heavily rewritten, sensationalized in fact, by John Conrad, who shared the byline. Martin the scholar was not happy with the tabloid conclusion to his investigations: no documentation, no referencing, no acknowledgments, and considerable groundless embellishment. However, Leo Rullman's challenge had been met, his ghost could rest, and Martin was very busy.

Some decades later, Martin was happy to share his documents, letters, and interview notes with Jeff Busby and Bart Whaley, who tracked down much more material, written, printed, and pictorial. Their book, *The Man Who Was Erdnase* [11], stands as a unique monument. Martin wrote in the Foreword:

> Bart Whaley, encouraged and assisted by Jeff Busby, has done a truly magnificent job of pulling together everything known today about Andrews and his masterpiece. He has set it all down in such loving detail, with such clarity, brilliance, and impeccable documentation, as to elevate him to the ranks of our country's top writers about true crime. I believe that this amazing book will become as famous in the literature of magic as Andrews' own classic. And what a sad, bitter, violent fantastic story it tells!

Martin contributed two articles, "The Mystery of Erdnase" (from the Program Book of the 1947 SAM Chicago Conference) and "The Man Who Was Erdnase," to *The Annotated Erdnase* by Darwin Ortiz [10], who added his comments—in all 11 pages of pertinent additional information.[2]

Returning to the year 1946, we find fortune smiled on Martin twice more. He began the first monthly column of his career, "Puzzles – Tricks – Fun," in *Uncle Ray's Magazine* for September 1946. Every year since, except 1982, when the Gardners moved from New York to North Carolina, Martin has been engaged in at least one monthly column. In the dizzy year of 1953, he ran six columns simultaneously: in *Hugard's Monthly, Polly Pigtails, PigglyWiggly, Humpty Dumpty's, Parents' Magazine*, and *Children's Digest*.

The year 1946 also marked Martin's first sale as a professional fiction writer. "The Horse on the Escalator" appeared in the October *Esquire*. Martin modeled the story's narrator after his early mentor and good friend, Dorny. In Martin's 1987 anthology, *The No-Sided Professor and Other Tales of Fantasy, Humor, Mystery, and Philosophy*, he wrote, "it was the sale of this story to *Esquire* that gave me the courage to decline an offer to have back my pre-

---

[2]The Erdnase mystery remains somewhat of a mystery. In an interview by Richard Hatch appearing in the April 2000 issue of *MAGIC*, Martin Gardner said, "you've convinced me now that there is good reason for doubt that Milton Franklin Andrews was Erdnase. I still think it was Milton Franklin, but my conviction rate is lowered ... to 60%."

war job in the press relations office of the University of Chicago. I wanted to see if I could earn a living as a writer."

New York, New York

Martin moved to New York in 1947 and rapidly entered the magic whirl: Saturdays at Lou Tannen's, then on to the restaurant or to Bruce Elliott's with Dai Vernon, Paul Curry, Clayton Rawson, Persi Diaconis, Bill Simon, Dr. Jaks, and/or other like-minded friends. Elliott's "The Back Room" column in *The Phoenix* provides a running account of New York activity: a line from #189 says, "managed to keep [Bill Woodfield] and Martin Gardner up till six A.M. which is considered par for the course in these parts."

During 1948, Martin initiated his ten-year run of monthly contributions to *Hugard's Monthly*, which became the basis of his mammoth *Encyclopedia of Impromptu Magic* (574 pages, 894 illustrations) [3]. Together with much new material, ideas, and references, there are about 2,000 items in 161 categories, from "Apples" to "Zipper," of unpredictable length: "Horn" has a paragraph, "Muscle Reading" has four pages, "Coins" has 138 entries, "Hands" has 96 entries. And, there are no card tricks and no rope tricks included! In his introduction, Martin continued to worry about the ideal format:

> I hoped that someday I might find time for extensive revisions and additions. I would obtain entry to a collector's library and plow through all his books, page by page. I would spend at least a few months on major periodicals. After that, I would attempt a comprehensive cross index.
>
> A trick, for instance, that uses a glass, handkerchief, and coin can be described only once, under one heading, but it should be cross referenced under other headings. Many tricks can be done with a variety of different objects. A trick with, say, a pencil may be equally effective with a table knife or a cane or a fountain pen. These, too, should be cross referenced as fully as possible; otherwise a reader searching the "Encyclopedia" for tricks with a certain object would be forced to go through the entire work if he wanted to run down all tricks applicable to that object. Also there should be cross indexing under such categories as "Practical Jokes," "Betchas," "Mental Effects," and so on, that would cut across listings by objects used.

In 1948, Martin also commenced writing on fringe science, cranks, impostors, cultists, and hoaxers. His first book for the public appeared in 1952, *In the Name of Science* [5]. A second edition, expanded to 363 pages, was issued by Dover in 1957 with the new name *Fads and Fallacies in the Name of Science*. Said Martin, "Don't care for the title myself, but the publisher wouldn't budge on it."

Karl Fulves published two booklets for the magic trade by Martin, under the name Uriah Fuller: *Confessions of a Psychic*, 1975, and *Further Confessions of a Psychic*, 1980. For general trade, Prometheus issued Martin's *How Not to Test a Psychic* in 1989. They also issued a companion volume to *Fads and Fallacies* in 1981, *Science: Good, Bad, and Bogus*, collecting Martin's articles and book reviews on pseudoscience and parapsychology up to 1981. Martin has been criticized for employing ridicule at times rather than reason, but he answers, "one horse laugh may be worth a thousand syllogisms."

The Committee for the Scientific Investigation of Claims of the Paranormal (CSICOP) was formed in 1976 with a journal, *The Skeptical Inquirer*. Martin's contributions are gathered by Prometheus in *The New Age: Notes of a Fringe-Watcher*, 1988, and *On the Wild Side*, 1992.

Martin burst out writing on several fronts in 1948: magic, philosophy, fiction, science. Only a few significant titles will be mentioned from his widening fields of endeavor. Returning to the strand of magic, in 1949 Martin linked his three homes neatly. His introduction to *Over the Coffee Cups* was dated, "New York, 1949" and was dedicated "To The Chicago Round Table Gang" and published by Montandan Magic in Tulsa. Friends gave Martin good leads on stories. In an article, "It Happened Even to Houdini," printed in *Argosy* for October 1950, for instance, he said:

> Dai Vernon, one of the greatest card magicians of all times, was performing his club act last summer on the Brazil, a steamship en route to Buenos Aires. Dai had a card selected, then placed it back in the deck. "When I throw this pack into the air," Dai said, "the chosen card will stick to the ceiling." Dai gave the deck a vigorous toss. To his great astonishment, the pack vanished completely! It had gone through a small ceiling air vent which he hadn't noticed because he'd been working under a spotlight.

In 1952, Martin entered two longterm relationships. Bill Simon had introduced Martin to Charlotte Greenwald, and now he served as best man at their wedding, performed by Judge George Starke, another magic friend. Also, Martin sold an article, "Logic

Machines," to *Scientific American*. Typically, Martin had a special page provided for the readers to cut up into logic window cards.

The year 1955 saw the birth of the Gardner's first son, Jim, and also the birth of a new book. "Fresh! Original! ... scores of new tricks, new insights, new demonstrations." For once, the blurbs were quite correct. Martin's *Mathematics, Magic and Mystery*, 1956, was loaded with great ideas! (The preface was dated 1955, but publication was delayed.) It contained 115 actions describing over 500 tricks—still in print and still inspiring [7]. Even sleight-of-hand experts Ed Marlo and Dai Vernon contributed, among a host of Martin's friends. The Gardners' son, Tom, was born in 1958.

Unknowingly, Martin reached a turning point in his career with the sale to *Scientific American* of a fascinating article on "Flexagons" for the December 1956 issue. These endlessly transforming paper foldovers were an immediate success. Could Martin produce a monthly column on mathematical games? Of course! Martin was off and running and didn't look back for 25 years!

His column became immensely popular as his topics broadened to include everything from art, through carnival swindles and computer games, to literature. Martin could not only explain abstruse scientific topics in ways intriguing to high school beginners, but he could also reveal unexpected depths in simple games and tricks, sufficient to challenge the professionals. In September 1977, Martin's "Mathematical Games" was moved from the back section to the first position in the front of *Scientific American*—a signal honor!

After 24 years of meeting monthly deadlines, Martin wrote only six columns in 1981, alternating months with his successor, D. R. Hofstadter. Martin finished with the December column, while Hofstadter kept the pace for 19 more columns, bowing out in July 1983.

Happily, Martin periodically gathered his columns into books, made even more interesting by added material and comments from readers. The first was *The Scientific American Book of Mathematical Puzzles and Diversions*, 1959 [9]. They grew in size over the years, with the 14th, *Fractal Music, Hypercards, and More*, 1992, having 327 pages [4]. These 14 volumes, totaling 3,829 pages, with a 15th forthcoming to complete the series, form an unparalleled source of classic concepts, current principles, and inspiration for new developments in magical and mathematical recreations.

Martin contributed three further "Mathematical Games" columns to *Scientific American*—August and September 1983, and June 1986—marking 35 years of association with the magazine.

Martin's most successful book (over 500,000 sold!) was published in 1960, *The Annotated Alice* (including both of Lewis Carroll's books *Alice's Adventures in Wonderland* and *Through the Looking Glass*) [2]. Many annotated works of all kinds had been produced in the past, but only for the scholarly community, with cramped footnotes, little grace, and no illustrations. Martin's book was large format with legible type, delightful style, and profuse illustrations. It was accepted by the general public, who are still buying and enjoying it. Thirty years later, Martin followed up with *More Annotated Alice* [8]. Martin annotated several other works, including *The Rhyme of the Ancient Mariner*, 1965; *Casey at the Bat*, 1967; and *The Night Before Christmas*, 1991. Other authors soon began "annotating" anything they could find in the public domain.

Martin had some fun in 1975! He wrote a straight-faced column (though scattered with clues) of fictitious "science developments" for the April issue (read: April Fool issue) of *Scientific American*. Material that would have drawn a good laugh in *Mad* magazine was treated with great respect! Some of the topics, complete with illustrations, were: Leonardi da Vinci's invention of the flush toilet, a fatal flaw in Einstein's theory of relativity, a map that required more than the usual four colors to complete, and a simple motor that ran on psychic energy. Martin received several thousand letters, most patiently pointing out the *one* error Martin had made in an otherwise excellent column!

Hendersonville, North Carolina

Martin's retirement from *Scientific American* at the end of 1981 prompted national attention and congratulations, with articles in *Newsweek*, *Omni*, and *Science 81*, among others. *Time* had written Martin up in 1975—the Mathemagician! He was credited with interesting more people in mathematics and science than anyone else alive. The quieter days in Hendersonville, North Carolina, allowed Martin to wrestle with the larger, intractable puzzles of life and the universe, of religion and society. Two books were soon readied.

In 1993, Martin gathered his separately marketed tricks and novelties and his original contributions scattered through dozens of books and magazine since 1930, revised and updated them, and added new material to form *Martin Gardner Presents* (424 pages, 230 tricks, 450 illustrations) [6]. To the wealth of material covering all small objects including cards and rope, Dana Richards,

Martin's official bibliographer, added an extremely useful "Bibliography of Martin Gardner in Magic" through the end of 1992, but including *Martin Gardner Presents*. This adds another member to Russell's self-inclusive class! Aside from books and pamphlets, Richard Kaufman records information very hard to find: book introductions, book reviews, articles not in magic magazines, marketed effects and puzzles, and 235 individual effects contributed to 28 different books and 28 (!) different magazines, from *Abracadabra* to *The Swindle Sheet*. The hundreds of tricks from Martin's series in Hugard's *Encyclopedia of Card Tricks* are not listed, since they are compiled in the *Encyclopedia of Impromptu Magic*.

In January 1993, a three-month puzzle exhibition opened in the Atlanta Museum of Art. Martin and Charlotte Gardner were honored guests at special "Gathering for Gardner" festivities, including the unveiling of the incredible portrait of Martin in dominoes.[3] A book was presented to each person there, *Martin, Articles in Tribute to Martin Gardner*, edited by Scott Kim. Of many good things therein, Dana Richards' "A Martin Gardner Bibliography" is outstanding. Sixty-eight pages of entries!

First, what it does not list: the *Scientific American* columns; magic tricks in magic periodicals; British editions and foreign translations; the individual stories, poems, and stunts in children's magazines.

Now, let us sample what is listed: 63 books and pamphlets; 19 books edited or annotated; 53 book introductions; 153 book reviews; 105 letters published; 210 columns and articles and anthologized material! Even 46 articles about Martin are listed! Dana Richards classified Martin's writing under 14 subject headings: Mathematics and Puzzles, Science, Fringe Science, Philosophy of Science, Philosophy and Theology, Political, Fiction, Poetry, Literature, Oziana, Juvenile Literature, Magic, Journalism, and Unclassified!

How to account for this amazing breadth of topics? How to sound the depths of Martin's creative talent? Truly, another "mysterious figure" is abroad, threading through our outposts, mugging a psychic here, skewering a charlatan there, spreading anti-irrational propaganda everywhere, then lightly dancing backward and away, confounding sober citizens with his laughter. When they ask, "Who are you, Nitram Rendrag?" he answers with G. K. Chesterton's riddle:

---

[3]An image of this domino portrait appeared on the cover of the April 1994 issue of *MAGIC* and can also be found on page ii of *The Mathemagician and Pied Puzzler*. [1]

I have a hat, but not to wear;
I wear a sword, but not to slay,
And ever in my bag I bear
A pack of cards, but not to play.

Bibliography

[1] Elwyn Berlekamp and Tom Rodgers. *The Mathemagician and Pied Puzzler*. Natick, MA: A K Peters, 1999.

[2] Martin Gardner. *The Annotated Alice*. New York: Potter, 1960.

[3] Martin Gardner. *Encyclopedia of Impromptu Magic*. Chicago: Magic, Inc., 1978.

[4] Martin Gardner. *Fractal Music, Hypercards, and More*. New York: W.H. Freeman, 1992.

[5] Martin Gardner. *In the Name of Science*. New York: Putnam, 1952.

[6] Martin Gardner. *Martin Gardner Presents*. New York: Kaufman and Greenberg, 1993.

[7] Martin Gardner. *Mathematics, Magic and Mystery*. New York: Dover, 1956.

[8] Martin Gardner. *More Annotated Alice*. New York: Random House, 1990.

[9] Martin Gardner. *The Scientific American Book of Mathematical Puzzles and Diversions*. New York: Simon and Schuster, 1959.

[10] Darwin Ortiz. *The Annotated Erdnase*. Pasadena, CA: Magical Publications, 1991.

[11] Bart Whaley with Martin Gardner and Jeff Busby. *The Man Who Was Erdnase*. Oakland, CA: Jeff Busby Magic, Inc., 1991.

Low-Down Triple Dealing

Colm Mulcahy

Consider the following three demonstrations of mathemagic:

1. A deck of cards is handed to a spectator, who is invited to
   shuffle freely. She is asked to call out her favorite ice-cream
   flavor; let's suppose she says, "Chocolate." Next, she is asked
   to cut off about a quarter of the deck and hold it ready for
   dealing.

   You take another quarter of the deck and demonstrate a spelling routine, dealing cards into a pile, one for each letter in
   the word "chocolate," before dropping the rest of your quarter
   deck on top. Set those cards aside and have the spectator
   perform this spelling routine three times with the cards in
   her hands. You correctly name the top card in her pile at the
   conclusion of her triple dealing.

2. A deck of cards is handed to two spectators, each of whom is
   invited to shuffle at will and then choose a card (of not too low
   a value) and place it face up on the table. Let's suppose that
   4♣ and 9♡ are selected and displayed. You run through the
   deck face up, tossing out all of the aces, twos, and threes—
   saying, "Sorry, I should have eliminated the low cards earlier."
   Then riffle shuffle a few times.

Remark, "Since a 9 was selected, let's count out nine cards," dealing into a pile on the table. Shuffle overhand and continue, "We'll need four more," as you peel off that many cards as a single unit, without changing their order. Drop these on top of the other nine. (The rest of the deck is ignored from now on.) Pick up this pile of thirteen cards and demonstrate dealing the nine top cards into a pile, thus reversing their order, and then dropping the remaining four cards on top. Have the first spectator do this deal three more times, and hand the cards to the second spectator. Have the second volunteer deal either four or nine cards into a pile, with the remaining cards placed beside this to form a second pile.

Recap: The two numbers (4 and 9) being used were determined by freely selected cards, and as a result, a deal of nine cards was performed (three times) on a packet of thirteen cards, which was then split into two piles. Draw attention to the two cards originally selected. Say, "Wouldn't it be surprising if, after all that triple dealing based on the values of two randomly selected cards from a shuffled deck, there were cards intimately related to the two you selected at the bottoms of the two piles now on the table?" Have the piles on the table turned over: one of the cards exposed is 9♣ and the other is 4♡. Add, "A curious alignment with the selected cards."

3. Have each of three volunteers in turn pick a card at random, and then have the cards returned to anywhere in the deck. Shuffle with abandon. Ask a fourth person to name their favorite magician, and assume "Harry Houdini" is called out. Hold the deck in the right hand, and peel cards off the bottom into a pile in the left hand, without altering their order, one for each letter, as you spell out the whole name. Hand the stack of twelve cards to the first volunteer and ask him to spell out HOUDINI while dealing out seven cards, then dropping the other five on top. Now give the cards to the second volunteer with the same directions, and finally to the third volunteer, for one last deal of the same type.

Take the cards behind your back and immediately produce three cards, handing one to each volunteer, face down. Have the chosen cards named, as they are turned over, to reveal that you have correctly located each one.

The same purely mathematical principle underlies each of these demonstrations, with a little more magic thrown in for good effect as we progress to the second and third tricks. We gradually reveal this principle below and discuss how each of the tricks is done as we go, before finally explaining why the principle works.

Let's start with the first effect. There are two secrets working behind the scenes for you here: an unadvertised but important relationship between the length of the word being spelled out and the size of the "quarter deck" with which the spectator starts, and the fact that you must somehow know the identity of one card in the spectator's hand from the beginning. It should come as no surprise that the card in question is the bottom card: asking the spectator to hold the cards in her hand in preparation for the spelling is just to give you an added opportunity to peek at this card, if you haven't already done that as she completed her shuffling. You must do whatever it takes to discover that card's identity!

This is the scoop on the ice-cream trick:

Claim 3.1. Start with n cards, the bottom one of which is known. If k cards are dealt out into a pile, thus reversing their order, and the remaining $n-k$ cards are dropped on top as a unit, and this type of deal is repeated twice more, then the known card rises like cream to the top—provided that $n \leq 2k$.

In the case of the 9-letter word CHOCOLATE, the trick works, provided that the portion of the deck selected by the volunteer contains at most 18 cards. If MINT CHOCOLATE CHIP (17 letters) is named, you'll ask for between a third and half of the deck. (If RUM is selected, try to force RUM RAISIN!)

The triple deal described is actually 75% of a rather interesting quadruple deal (which can be used as the basis for an in-hand false shuffle).

This is the real scoop:

Claim 3.2. Start with n cards, and assume that $n \leq 2k \leq 2n$. If k cards are dealt out into a pile, thus reversing their order, and the remaining $n-k$ cards are dropped on top as a unit, and this deal is repeated three more times, the entire packet of n cards is restored to its original order.

Now consider the second effect. Two cards (not aces, twos, or threes) are chosen and set aside face up. Let's suppose that they are 4♣ and 9♡. As you run through the deck face up, ostensibly to toss out the low valued cards, what you really focus on doing is cutting the 9♣ and 4♡ to the top and bottom, respectively. They will stay there if you are careful how you riffle shuffle. Continue as described earlier: reversing nine cards into a pile and then doing some overhand shuffling, whose purpose is to bring the bottom card to the top. Peel four more off the top without reversing them, and drop on top of the other nine. You now have thirteen cards, with the desired two cards at the top and bottom of that packet. Your subsequent demonstration of dealing nine and dropping four is just the first of a series of four deals: the first spectator does the next three deals, thereby restoring the packet to its initial state. The second spectator deals (either four or nine) cards into a pile, and then there are two piles on the table, with one of the desired cards at the bottom of each pile. You are all set for the grand finale.

The third effect uses the fact that after three deals of the type described, not only does the bottom card rise to the top, but the next to last card becomes the second card from the top, the one above that becomes the third card, and so on.

This is the real triple scoop:

Claim 3.3. If k cards from n cards are dealt out into a pile, reversing their order, and the remaining $n - k$ are dropped on top as a unit, and this process is repeated twice more, then provided that $n \leq 2k \leq 2n$, the original k bottom cards become the top k cards, in reverse order.

To perform the third trick, ask each of three volunteers to pick a card at random. Have these cards returned, one at a time, to the deck and *then control them to the bottom*—this means that you appear to allow free choice of where to put the cards, but you actually use elementary magic techniques to get each card to the bottom. As a result, the third volunteer's card is at the bottom of the deck, the second volunteer's card is one up from the bottom, and the first volunteer's card is two up from the bottom. Peel cards off the bottom of the deck—without altering their order—one for each letter of the name of the magician called out, as you spell out both words in full. Hand the resulting packet of cards to the first volunteer and ask that the longer of the two names (HOUDINI in our example) be spelled out as cards are dealt into a pile, before dropping the remainder on top. Now give the cards to the second volunteer and finally to the third volunteer for two more deals. The three chosen cards are now on the top of the packet of cards, with the order reversed, and you are all set to conclude in triumph.

Why are all of the above claims valid for any n and k with $n \leq 2k \leq 2n$? It's certainly easy to see if $n = k$ (reversing all of the cards

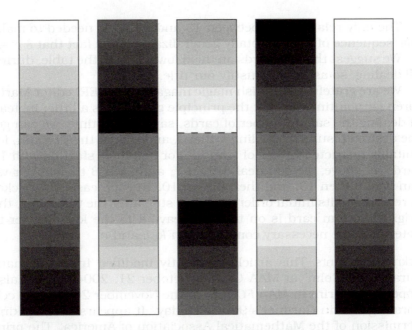

Figure 1. Proof without words. The dashed lines indicate the portions that move around intact, subject at most to some internal reversals.

each time), and almost as easy to see if $n = 2k$ (reversing exactly half of the cards). Actually, it's easy to *see* in all cases.

Suppose for the sake of concreteness that $n = 13$ and $k = 8$. Let's represent a pile of thirteen cards in some initial order by a vertical strip of gray-scale panels in decreasing order of brightness, from white for the top card to black for the bottom card, as depicted in the leftmost strip in Figure 1.

Then, the results of the four deals—each of eight cards into a pile with the other five cards dropped on top—is given by the successive vertical strips. Since the rightmost strip shows a fully restored pile, the deal in question has period four: after four deals, we are always back to where we started. After three such deals, the original bottom card (black) has risen to the top—in preparation for its final journey back to the bottom under one more deal. Moreover, it is clear that the eight bottom cards become the eight top cards, suitably reversed, after three deals. There are just three portions of the packet—of sizes five, three, and five here—of which to keep track, and they move around intact, subject at most to some internal reversals (indicated by the dashed lines in Figure 1).

The only relationship between 13 and 8 that is needed to make this sequence of images totally generalizable is the fact that $8 \geq \frac{13}{2}$.

We suggest that the cards are held low, close to the table, during all dealing, so as to fully justify our title.

We are grateful to Finnish magic magazine *JOKERI* editor Martti Sirén for pointing out that the principle generalizes a little: instead of dealing the same number of cards, say k, each time, we can get the desired results by dealing k then l, and then k then l again, for suitably restricted values of k and l. For instance, starting with 13 cards as above, we can deal 8 then 7, and then 8 then 7. Or we can deal 8 then 10, and then 8 then 10. In each case, the packet is restored to its initial order and if we stop after the third deal, the original bottom card is on top. We leave it to the keen reader to determine the necessary conditions on k, l, and n.

Acknowledgments. This article is slightly modified from the inaugural "Card Colm" at MAA Online (October 21, 2004), which also appeared in print in *MAA FOCUS*, in the November 2004 issue celebrating Martin Gardner's 90th birthday. It appears here by kind permission of the Mathematical Association of America. The principle involved was discovered, while on leave, in a Madrid suburb in spring 2003. More mathematical card tricks can be found at http://www.spelman.edu/~colm/cards.html.

A Lifetime of Puzzles

Products of Universal Cycles

Persi Diaconis
Ron Graham

Universal cycles are generalizations of de Bruijn cycles to combinatorial patterns other than binary strings. We show how to construct a *product cycle* of two universal cycles, where the window widths of the two cycles may be different. This mathematical theorem leads to applications in card tricks.

Introduction

A *de Bruijn cycle* is a sequence of zeroes and ones such that each window of width k running along the sequence shows a different binary k-tuple. We assume the ends of the sequence are joined to form a cycle. For example, when $k = 3$,

$$10111000$$

shows $101, 011, 111, 110, 100, 000, 001, 010$ (where the window is run "around the corner"). In this example, the total length of the cycle is $2^3 = 8$, so each 3-tuple appears once. This is not required, so 000111 (for $k = 3$) is a de Bruijn cycle of length six. The reader may enjoy the problem of constructing a de Bruijn cycle of length 52 for a window of width $k = 6$.

De Bruijn cycles are used for cryptography, robot vision, random number generation, and DNA sequencing. We show elsewhere that they can be used secretly for card tricks [2, Chapters 2–4]. There are also a friendly introduction to de Bruijn sequences by Stein [15, Chapter 8], a more comprehensive survey by Fredericksen [4], and an extensive discussion by Knuth [13, Section 7.2.1.1]. These articles show that maximal-length de Bruijn sequences always exist for any window width k. They give a variety of constructions and properties (for example, they show how to determine the binary pattern in position t). It is even known exactly how many maximal-length de Bruijn sequences there are: $2^{2^{k-1}-k}$.

In joint work with Fan Chung [1], we have introduced a generalization called *universal cycles* that extends the notion from binary strings to other combinatorial pattern, such as the relative order of k consecutive symbols. Thus, consider for window width 3 the sequence

$$1\,3\,2\,1\,3\,4.$$

The relative order of the first three numbers is *low-high-medium* or LHM. The successive relative orders (going around the corner) are

LHM, HML, MLH, LMH, MHL, HLM.

Thus, each of the six possible relative orders (or permutations) appears exactly once. This is an example of a universal cycle for permutations. In our work with Chung, we showed that, for every k, the numbers $1, 2, 3, \ldots, k!$ can be arranged so that each consecutive block of k has a distinct relative order. While such sequences were shown to exist, no general rule for construction, nor any formula (or approximation) for the total number is known. The reader may enjoy one of the following two problems:

- Write the numbers $1, 2, 3, \ldots, 24$ in a sequence so that each successive group of *four* shows a distinct relative order.

- Write down a sequence of length 24 using only the numbers $1, 2, 3, 4, 5$ so that each successive group of *four* shows a distinct relative order.

We have also constructed sequences of symbols $1, 2, \ldots, n$ so that each consecutive k-tuple shows a distinct k-subset from the set $\{1, 2, \ldots, n\}$. These only exist for certain k and n; even the existence is an open research problem. This is a universal cycle for k-subsets of an n-set. (See [8, 11] for more details.)

More generally, given any natural combinatorial object describ-ed by k parameters $(\theta_1, \theta_2, \ldots, \theta_k)$, one may ask for a sequence of θ-values so that each consecutive block of k codes exactly one of our objects. More carefully, there is a fixed finite alphabet Θ, and each θ_i is in Θ. Further, there is a rule $R(\theta_1, \theta_2, \ldots, \theta_k)$ taking values *one* or *zero*. Our combinatorial object is the set of all $(\theta_1, \theta_2, \ldots, \theta_k)$ so that $R(\theta_1, \theta_2, \ldots, \theta_k) = 1$. For example, if $\Theta = \{1, 2, \ldots, k\}$ and $R(\theta_1, \theta_2, \ldots, \theta_k)$ is one if the θ_i are distinct, and zero otherwise, then the combinatorial object is the set of all permutations on k symbols.

A variety of constructions have appeared: set partitions, or-dered k-out-of-n, subspaces of a vector space, and others. It seems fair to say that, up to now, the construction of universal cycles has proceeded by clever, hard, ad-hoc arguments. There is nothing like a general theory.

The purpose of the present article is to begin a theory by show-ing that, for some cases, *products* of universal cycles can be formed. In the next section, we introduce the product construction by tak-ing the product of 10111000 and 132134. A card trick version is given, along with a general recipe for the product of a de Bruijn cycle and an arbitrary universal cycle—both with the same window width k. The following section gives products for universal cycles more general than de Bruijn cycles with an arbitrary universal cy-cle, which have (possibly) differing window widths. Following this is a practical section that concerns "cutting down" universal cycles (e.g., from 64 to 52). Proofs are deferred to the appendix, which gives a very general product construction.

Products with Equal Window Widths

Suppose that $x_1 x_2 \ldots x_R$ and $y_1 y_2 \ldots y_S$ are each universal cycles for the same window width k. We want to use these cycles to form a sequence of *pairs*

$$
\begin{array}{cccc}
x_1 & x_2 & \ldots & x_{RS} \\
y_1 & y_2 & \ldots & y_{RS}
\end{array}
$$

so that a window of width k, run along the pairs, shows each of the possible (vertical) pairs of x-tuples and y-tuples just once. The eas-iest case occurs when the integers R and S have no common factor larger than 1. We may then simply write $x_1 x_2 \ldots x_R x_1 x_2 \ldots x_R \ldots$ $x_1 x_2 \ldots x_R$, repeated S times. Under this row, write $y_1 y_2 \ldots y_S y_1 y_2 \ldots$ $y_S \ldots y_1 y_2 \ldots y_S$ repeated R times.

Example 1. Suppose that $R = 3$, $S = 4$, and $k = 2$. Let the x sequence code the relative order with ties permitted. Thus, two successive values may be *low-high* (LH) or *high-low* (HL) or *equal* (EQ). Thus, the sequence 112 gives EQ, LH, HL (where the last pair comes from going around the corner). The y sequence uses zeroes and ones, such as 0011, for the usual de Bruijn sequence for window width $k = 2$. Here, $RS = 12$. The product is

$$
\begin{array}{cccccccccccc}
1 & 1 & 2 & 1 & 1 & 2 & 1 & 1 & 2 & 1 & 1 & 2 \\
0 & 0 & 1 & 1 & 0 & 0 & 1 & 1 & 0 & 0 & 1 & 1.
\end{array}
$$

Our interest in forming products arose from a card trick. We wanted to take a product of the usual de Bruijn sequence 10111000 of length eight with the permutation sequence 132134 of length six. Both have window width $k = 3$. This would give an arrangement of 48 cards, so that the relative order and color pattern of successive triples uniquely determines the position. We originally constructed an example in an ad-hoc fashion (the naïve construction above doesn't work because $R = 6$ and $S = 8$ are not relatively prime). We then developed some theory (described below). The following describes the construction that the theory gives.

Take an ordinary deck of cards. Remove the four kings. Arrange the rest in the following order (ace is low), with D, C, H, and S for diamonds, clubs, hearts, and spades, respectively:

A	7	5	3	9	J	2	7	6	3	10	Q	A	8	5	4
D	C	D	H	H	S	C	S	H	S	D	H	H	C	S	C

10	Q	2	7	5	3	10	J	A	8	6	4	10	J	2	8
H	S	D	D	H	C	S	H	C	H	D	D	C	C	S	D

6	4	9	J	A	8	6	3	9	Q	2	7	5	4	9	Q
C	H	D	D	S	S	S	D	C	D	H	H	C	S	S	C.

The reader will find that each successive group of three cards is uniquely identified by the relative order and color pattern. For example, the top three cards are AD, $7C$, $5D$ and have relative order *low-medium-high* (LHM) and color pattern *red-black-red* (RBR). No other successive triple has both of these patterns. This property can be used to perform a card trick. Put the 48-card deck, arranged as above, in the card case. Find an audience—the larger, the better. Have the audience members take the cased deck and pass it to the back of the hall. Have a "randomly chosen spectator" cut the deck and complete the cut. The deck is passed to a second, adjacent, spectator, who also gives the cards a complete cut. Then, the deck is passed to a third spectator, who gives it a complete cut. This third spectator removes the current top card, showing it to no one. The deck is passed back to the second spectator, who removes the top card and then passes the deck back to the first spectator who removes the top card. The performer patters as follows:

> Three of you have freely cut the cards and selected a card. I'd like you to look at your card and concentrate, form a mental picture, and try to project. You're doing a great job, but it's hard to unscramble things. Let me try this. I see red more clearly than black. Would everyone with a red card please stand up? That helps a lot! But still, it's not in focus. Who has the highest card of you three? [One of the spectators waves.] Who has the lowest? [Again, one of the spectators waves.] I'll work on the middle man first. You, sir, have a spade... it's the nine of spades? Now, the high man. You have a high black card. Is it the queen of clubs? Finally, the lady with the lowest card. It's a four. Is it the four of spades?

Practical performance details for tricks of this type are in [2, Chapter 2]. We mention the magic application to explain our motivation for the constructions in the present paper.

The 48-card arrangement was constructed in two stages. First, we used the product theorem described below to get a "product" of the universal cycle for permutations (namely, 132134) with the de Bruijn cycle (namely, 10111000). This results in the following sequence of 48 pairs:

$$
\begin{array}{cccccccccccccccccccccccc}
1 & 3 & 2 & 1 & 3 & 4 & 1 & 3 & 2 & 1 & 3 & 4 & 1 & 3 & 2 & 1 & 3 & 4 & 1 & 3 & 2 & 1 & 3 & 4 \\
1 & 0 & 1 & 1 & 1 & 0 & 0 & 0 & 1 & 0 & 1 & 1 & 1 & 0 & 0 & 0 & 1 & 0 & 1 & 1 & 1 & 0 & 0 & 1 \\
\end{array}
$$

$$
\begin{array}{cccccccccccccccccccccccc}
1 & 3 & 2 & 1 & 3 & 4 & 1 & 3 & 2 & 1 & 3 & 4 & 1 & 3 & 2 & 1 & 3 & 4 & 1 & 3 & 2 & 1 & 3 & 4 \\
0 & 1 & 1 & 1 & 0 & 0 & 0 & 1 & 0 & 1 & 1 & 1 & 0 & 0 & 0 & 1 & 0 & 1 & 1 & 1 & 0 & 0 & 0 & 0. \\
\end{array}
$$

Notice that the bottom row is not quite a repetition of 10111000. Nonetheless, the product theorem guarantees that each three successive pairs are uniquely determined by the relative order of the top sequence and the zero–one pattern of the bottom sequence.

The second stage requires lifting the last pattern to the natural context of playing cards, with twelve values, each repeated four times, and four suits. We explain how to lift below. We now state a first product theorem.

Theorem 4.1. (Product of a de Bruijn and universal cycle with equal window lengths) *Let $\overline{x} = x_1 x_2 \ldots x_R$ be an arbitrary universal cycle* with window width k. Let $\bar{y} = y_1 y_2 \ldots y_S$ be a de Bruijn cycle with window width k. Here, the symbols x_i can be in any alphabet, and the symbols y_j are zero–one. Neither cycle need be maximal, but we do assume that $\bar{y}$ ends with k consecutive zeroes. The following construction gives a sequence of pairs

$$x_i$$
$$y_j$$

of length RS so that, if a window of width k is run along the pairs, each ordered k-tuple of x_i's above an ordered k-tuple of y_j's appears just once.

> If the sequence lengths R and S have no common factor, repeat the $\bar{x}$ sequence S times above the $\bar{y}$ sequence repeated R times. If the largest number that divides R and S is d, write $R = rd$ and $S = sd$. Observe that r and s are relatively prime. Begin by writing down the $\bar{x}$ sequence S times, forming a sequence of length RS. Under this, we construct the following sequence. Recall that $\bar{y}$ is a de Bruijn sequence with k zeroes at the end. Form a string of zeroes and ones by repeating the original sequence $\bar{y}$ a total of r times, and then removing the final zero. This gives a sequence $\bar{y}^*$ of length $rS - 1$. Now, repeat the $\bar{y}^*$ sequence d times, and finish off with a string of d zeroes. Place this, in order, under the $\bar{x}$ sequence of length RS.

Example 2. Look back at the product of $1\,3\,2\,1\,3\,4$ and $1\,0\,1\,1\,1\,0\,0\,0$ in the preceding section. Here, $R = 6$, $S = 8$, and so the greatest divisor is $d = 2$. The construction begins by repeating $1\,3\,2\,1\,3\,4$ eight times to form the top row. For the bottom row, $r = 3$. The block $\bar{y}^*$ is formed from three copies of $1\,0\,1\,1\,1\,0\,0\,0$ and then deleting the final zero. Thus, $\bar{y}^* = 1\,0\,1\,1\,1\,0\,0\,0\,1\,0\,1\,1\,1\,0\,0\,0\,1\,0\,1\,1\,1\,0\,0$. The bottom row is formed from $d = 2$ copies of $\bar{y}^*$ followed by $d = 2$ further zeroes.

Example 3. Let us take the product of $1\,1\,0\,0$ with itself. Thus, $R = S = d = 4$ and $r = s = 1$. The top sequence is formed from using four copies of $1\,1\,0\,0$. For the bottom row, the building block is $\bar{y}^* = 1\,1\,0$. Placing four repetitions of this followed by four final zeroes gives the product sequence

$$1\ 1\ 0\ 0\ 1\ 1\ 0\ 0\ 1\ 1\ 0\ 0\ 1\ 1\ 0\ 0$$
$$1\ 1\ 0\ 1\ 1\ 0\ 1\ 1\ 0\ 1\ 1\ 0\ 0\ 0\ 0\ 0.$$

The reader may check that the sixteen 2×2 windows are distinct:

$$
\begin{array}{cc@{\qquad}cc@{\qquad}cc@{\qquad}cc@{\qquad}cc@{\qquad}cc@{\qquad}cc@{\qquad}cc}
1 & 1 & 1 & 0 & 0 & 0 & 0 & 1 & 1 & 1 & 1 & 0 & 0 & 0 & 0 & 1 \\
1 & 1 & 1 & 0 & 0 & 1 & 1 & 1 & 1 & 0 & 0 & 1 & 1 & 1 & 1 & 0 \\[6pt]
1 & 1 & 1 & 0 & 0 & 0 & 0 & 1 & 1 & 1 & 1 & 0 & 0 & 0 & 0 & 1 \\
0 & 1 & 1 & 1 & 1 & 0 & 0 & 0 & 0 & 0 & 0 & 0 & 0 & 0 & 0 & 1.
\end{array}
$$

Example 4. Of course, our product construction can be iterated. Consider the extreme case of $k = 1$. A de Bruijn cycle for $k = 1$ is 10. The product of this with itself (using the product theorem) is

$$
\begin{array}{cccc}
1 & 0 & 1 & 0 \\
1 & 1 & 0 & 0.
\end{array}
$$

We may now take the product of this with 10 to get

$$
\begin{array}{cccccccc}
1 & 0 & 1 & 0 & 1 & 0 & 1 & 0 \\
1 & 1 & 0 & 0 & 1 & 1 & 0 & 0 \\
1 & 0 & 1 & 1 & 0 & 1 & 0 & 0.
\end{array}
$$

Another product with 10 gives

$$
\begin{array}{cccccccccccccccc}
1 & 0 & 1 & 0 & 1 & 0 & 1 & 0 & 1 & 0 & 1 & 0 & 1 & 0 & 1 & 0 \\
1 & 1 & 0 & 0 & 1 & 1 & 0 & 0 & 1 & 1 & 0 & 0 & 1 & 1 & 0 & 0 \\
1 & 0 & 1 & 1 & 0 & 1 & 0 & 0 & 1 & 0 & 1 & 1 & 0 & 1 & 0 & 0 \\
1 & 0 & 1 & 0 & 1 & 0 & 1 & 1 & 0 & 1 & 0 & 1 & 0 & 1 & 0 & 0.
\end{array}
$$

Can the reader see the simple pattern, and how it will continue? (*Hint:* read the columns upside-down and backwards, in binary.)

We conclude this section with a few remarks that extend the construction in various ways. The construction given was for $\bar{y}$, a zero–one de Bruijn sequence. This is not at all required. The construction works for *any* universal cycle with window width k that contains a block of k repeated symbols. We will call such universal cycles *special.* Here are several other such examples.

Example 5. We have given a universal cycle for partitions of an n-element set [1]. These partitions are counted by the Bell numbers. Martin Gardner gives a wonderful introduction to these numbers [5, Chapter 2] (see also [14, Section 7.2.1.5] for some extensions). For example, there are 15 partitions of the four-element set $\{1, 2, 3, 4\}$:

1234	1|234	12|34	1|2|34	1|2|3|4.
	2|134	13|24	1|3|24	
	3|124	14|23	1|4|23	
	4|123		2|3|14	
			2|4|13	
			3|4|12	

A universal cycle for these is $1\,2\,3\,2\,3\,3\,3\,3\,4\,4\,3\,4\,5\,5\,3$. As a window of width four is run along, the equal positions run through all possible set partitions exactly once (so that $1\,2\,3\,2$ corresponds to the partition $1|3|24$, $2\,3\,2\,3$ corresponds to the partition $13|24$, etc.). Because there is a block of four repeated symbols, namely $3\,3\,3\,3$, these can be cycled to the end, and then the product with any universal cycle can be formed.

Example 6. We have given a universal cycle for permutations with ties [3]. For example, there are 13 possible relative orders of three distinct values when ties are allowed. They are

123, 132, 213, 231, 312, 321, 112, 121, 211, 221, 212, 122, 111.

A universal cycle for these permutations of three symbols with ties (using the symbols $\{1, 2, 3, 4\}$) is given by $1\,1\,1\,2\,1\,2\,2\,1\,3\,4\,1\,3\,2$. For any window width k for these universal cycles, there is always a block of k repeated symbols (all k values are tied), so these cycles are special and can be used in the product theorem.

Alas, not every universal cycle has a block of repeated symbols, and we are at a loss for a general construction. For example, when $k = 3$, with the permutation cycle $1\,3\,2\,1\,3\,4$, we do not know how to form a product of $1\,3\,2\,1\,3\,4$ with itself. However, we can construct a cycle of 36 pairs of numbers so that the relative order of the top and bottom blocks of three are all distinct. We just need a little more freedom in the alphabet size. An example of such a cycle is

1	3	2	1	3	4	1	3	2	1	3	4	1	3	2	1	3	4
1	4	3	2	5	1	4	3	2	5	1	4	3	2	5	1	4	3

1	3	2	1	3	4	1	3	2	1	3	4	1	3	2	1	3	4
2	5	1	4	3	2	5	1	4	3	2	5	6	7	8	9	10	11.

For this sequence, the first block of three is

1 3 2
1 4 3.

The top row is in order L H M, and the bottom row is in the same order L H M. This is the only time this pair of orders appears together. Similarly, every pair of blocks of three (going around the corner) has a unique signature.

We created this sequence by "lifting" (see the section "Lifting/ Lumping" below) 1 3 2 1 3 4 to have all distinct symbols, e.g., 1 4 3 2 5 6. Here, every block of three has a unique relative order. This lifting has the property that it can be "cut down" (see the section "Cutting Down" below) to 1 4 3 2 5. This has every block of three with distinct relative orders, but omits L M H. Pasting six copies of this under six copies of our original 1 3 2 1 3 4 leaves six places to fill at the end. We filled them with 6 7 8 9 10 11 to give L M H six times, with all possible order parameters occurring on top. What is also crucial (and this is the result of "fooling around" and not of "higher math") is that the construction works going around the corner. Thus, the last block of two,

$$3 \quad 4$$
$$10 \quad 11,$$

combined with the first block of length one,

$$1$$
$$1,$$

gives

$$3 \quad 4 \quad 1$$
$$10 \quad 11 \quad 1.$$

On top we have M H L and on the bottom M H L. This is the only time this occurs. Similarly,

$$4 \quad 1 \quad 3$$
$$11 \quad 1 \quad 4$$

gives H L M and H L M uniquely. We hope that explaining our construction demystifies it and suggests further ideas for progress in constructing products.

More General Products

In the section above, we have explained how to form a product of a universal cycle with a de Bruijn cycle *provided* that the two cycles have the same window width k. It turns out that the construction given works even if the window widths are different.

Theorem 4.2. *The construction given in Theorem 4.1 works* mutatis mutandis *for a universal cycle with variable window lengths.*

As explained below, this generalization can be applied to card tricks in at least two ways. Suppose that $\bar{x} = x_1 x_2 \ldots x_R$ and $\bar{y} = y_1 y_2 \ldots y_S$ are universal cycles of respective window widths k and l. Suppose that we can construct a sequence of pairs

$$
\begin{array}{cccc}
x_1 & x_2 & \cdots & x_{RS} \\
y_1 & y_2 & \cdots & y_{RS}
\end{array}
$$

with x_i in the alphabet used for the $\bar{x}$ sequence and y_j in the alphabet used for the $\bar{y}$ sequence. The construction is a *product* if, starting at any position

$$
\frac{x_i}{y_i},
$$

the symbols $x_i x_{i+1} \ldots x_{i+k-1}$ and $y_i y_{i+1} \ldots y_{i+l-1}$ uniquely identify i.

Here is a simple example. Take $\bar{x} = 1\,3\,2\,1\,3\,4$. With $k = 3$, this is a universal cycle for permutations. Take $\bar{y} = RRWBBRBWW$. With $l = 2$, successive windows of width 2 go through each of the nine ordered pairs of "colors" {red, white, blue} or $\{R, W, B\}$. The lengths $R = 6$ and $S = 9$ have the largest common factor of $d = 3$. We may follow the construction of the product theorem virtually word for word. Build a sequence of total length $RS = 54$ by first repeating S copies of the $\bar{x}$ sequence. For the second row, $R = rd = 2 \cdot 3$. Repeat the $\bar{y}$ sequence $r = 2$ times and delete the final symbol. This gives $\bar{y}^* = RRWBBRBWWRRWBBRBW$. Next, repeat the $\bar{y}^*$ sequence d times and finish off with d repetitions of the deleted symbol W. This gives the final construction:

1	3	2	1	3	4	1	3	2	1	3	4	1	3	2	1	3	4
R	R	W	B	B	R	B	W	W	R	R	W	B	B	R	B	W	R

1	3	2	1	3	4	1	3	2	1	3	4	1	3	2	1	3	4
R	W	B	B	R	B	W	W	R	R	W	B	B	R	B	W	R	R

1	3	2	1	3	4	1	3	2	1	3	4	1	3	2	1	3	4
W	B	B	R	B	W	W	R	R	W	B	B	R	B	W	W	W	W.

Consider this sequence. The relative order of the first three places is LHM. The colors of the first two places are R. The reader may check that this pattern only occurs at the start: The RR pattern occurs just six times, and the three symbols directly above occur in six distinct relative orders.

Here are two ways such a product could be used for a card trick. First, suppose $k \geq l$. Have the deck of RS cards cut freely and have k consecutive cards taken off. Each card is representated by a labeled

$$x_i$$
$$y_i$$

pair. Ask for the $\overline{x}$ information. Then, ask the first l people for the $\overline{y}$ information. This combined information uniquely identifies the k cards. Here is a second procedure: Have $k + l$ cards taken off. Ask the first k people for the $\overline{x}$ information and the last l people for the $\overline{y}$ information. This information uniquely specifies the cards.

As a check, the reader may try both procedures out on the sequence of length 54 given above.

Combining techniques, we have a way of taking a product of two universal cycles, one of window width k and one of window width l, provided only that at least one of the two cycles contains a window of all identical symbols (or that the lengths of the two cycles are relatively prime). Neither cycle need be maximal or de Bruijn. We call this the *General Product Construction* (Theorem 3). The validity of the construction described, and somewhat more, will be proved in the appendix. Of course, higher-order products can be constructed by iterating the procedure.

Some Practical Details and Problems

Over the years of working with de Bruijn sequences and their generalizations, four practical problems have emerged. These are

- products,
- cutting down,
- lifting/lumping,
- coding and neat generation.

We have treated products earlier. We briefly treat the three remaining problems here.

Cutting Down

Maximal-length universal cycles come in tightly circumscribed lengths. For binary de Bruijn sequences, these lengths are 2, 4, 8, 16, 32, 64, ... Practical considerations may demand a shorter total length. For example, 52 is boxed between 32 and 64. We know there are maximal-length de Bruijn sequences of length 32 (window width 5) and 64 (window length 6). What do we know for 52? One approach would be to take a de Bruijn sequence of length 64 and cut it down. Remove segments of total length 12 so that the remaining 52 show all distinct patterns when a window of width six is run along the cycle. With patience, the reader will find that this is indeed possible.

However, there is a very beautiful method for accomplishing this for *any* cycle length in *any* longer maximal de Bruijn cycle. While we would love to take credit for this, it actually belongs to the "folklore" of de Bruijn cycles. (We thank Hal Fredericksen and Al Hales for tracking this down for us.) We have it catalogued as "Babai's Cutting Down Lemma." It gives a way of taking a de Bruijn sequence created from a "shift register" of total length $2^k - 1$ (window width k) and cutting down to *any* length L, $k \leq L \leq 2^k - 1$.

We will illustrate this method with the example $k = 6$, $2^k - 1 = 63$, and $L = 52$. First, consider the long de Bruijn sequence (length 63)

0000010000|1100010100|1111010001|

1100100101|1011101100|1101010111|111

where we have put |'s after every ten symbols to help count. Label the positions starting at the extreme left (position zero) to the extreme right (position 62). For example, the window of width six that starts at position 38 is 011011. The window starting at position 49 is 011010. These will figure into the discussion in a moment. As with any de Bruijn sequence, a window of width six run along the sequence shows all distinct blocks (except 000000, which does not occur in sequences generated by shift registers).

The sequence is formed from the starting block 000001 by a simple rule: form the symbol at position $i + 6$ by adding the two symbols at positions i and $i + 1$ (mod 2). Any of the articles cited in our references show how to find such rules, and their relation to the algebraic fact that $1 + x + x^6$ is a "primitive" polynomial.

To get from 63 to 52, we must remove 11 symbols. Note that $49 - 38 = 11$ and that the blocks starting at 38 and 49 differ in their last symbol only. We may thus simply replace the last symbol (a 1) in the 38 block 0 1 1 0 1 1 by a zero, cut out the intervening symbols, and get

0000010000|1100010100|1111010001|1100100101|1010101111|11.

The point of this construction (other than that it works) is this: the cut-down sequence of length 52 is *still* generated by a simple rule. Consider any block of length six. The next symbol after a block of six is formed by adding the first two symbols of the block (mod 2) *unless* the string formed from symbols two through seven would become $0\,1\,1\,0\,1\,1$. In that case, and in that case only, the simple rule is broken and a zero is adjoined instead, forming the block $0\,1\,1\,0\,1\,0$. This can happen at most once.

We claim that, for any window width k and any cut-down size, the same scheme works, with just one forbidden sequence. To see why (briefly), let $P(x)$ be the primitive polynomial generating your de Bruijn sequence. Our references show that the symbols are generated by raising x to successive powers $x, x^2, x^3, x^4, \dots$ (mod $P(x)$). If we seek to cut out a chunk of length c (in our case, $c = 11$), we need a power i so that

$$x^{i+c} + x^i \equiv 1 \pmod{P(x)} \quad \text{or} \quad x^i(x^c + 1) \equiv 1 \pmod{P(x)}.$$

Because the nonzero elements mod $P(x)$ form a field, this equation always has a unique solution for i, as long as $c \neq 2^k$. The string coded by x^i begins at 38 in our example. All of this will be Greek to the reader who does now know about finite fields; we hope it induces some to learn more!

We make two final remarks. First, finding the i that works must be done by trial and error (it is equivalent to finding logarithms in finite fields). Second, the procedure can be written as a simple "nonlinear" recurrence. In the example,

$$x_{n+6} \equiv x_n + x_{n+1} + \bar{x}_n x_{n+1} x_{n+2} \bar{x}_{n+3} x_{n+4} x_{n+5} \pmod{2},$$

where $\bar{x}_j$ denotes $1 - x_j$. Indeed, the nonlinear term is always zero, except at 011011 when it is one, shifting the sequence. We find this a truly beautiful application of mathematics to solve a practical magic problem (see [6, Section 7.5] for further details).

The same problem arises for other universal sequences. Here is a practical example. A Tarot deck consists of 78 cards; there are four suits, each of 14 values for 56 "ordinary" cards, and 22 cards in the "major arcana." These are trump cards with colorful names such as "The Hanged Man." Tarot cards have been around for over 500 years (see the marvelous history of Stuart Hampshire [7]). They are frequently used for fortune telling, and it is natural to try to invent one of our card tricks using Tarot cards. We use them here as an excuse for discussing available techniques.

One simple approach is to set aside the major arcana and work with the remaining 56 cards. Now $56 = 8 \cdot 7$. An easy construction is to take the usual de Bruijn sequence $1\,0\,1\,1\,1\,0\,0\,0$ for eight and then cut it down to seven as 1011100. Taking a product of these two will do the job.

Going back to the full deck of $78 = 6 \cdot 13$, there is a natural universal cycle of length 13 (permutations with ties with window width 3 given in Example 6), as well as one of length 6 (permutations with window width 3). In this case, it is possible to form a product universal cycle, because the cycle for permutations with ties does indeed have a block of three repeated symbols. Even easier in this case is to use the fact that the two cycle lengths 13 and 6 are relatively prime. However, a general theory is lacking on how to do this when a sufficiently long block of repeated symbols does not occur in either sequence in the product. To crystallize things, we state the situation as an open problem.

Problem 1. Let $x_1, x_2, \ldots, x_R$ be a universal cycle with window width k. For $k \leq j \leq R$, is it always possible to find a subsequence of length j that is also a universal cycle with window width k?

Lifting/Lumping

Lifting involves resizing a universal cycle based on a small alphabet with a larger alphabet. *Lumping* involves the opposite. Both are well illustrated with universal cycles for permutations.

First, consider lifting. In the second section, we took the product of the permutation cycle 132134 with the binary cycle 10111000. Both have window width $k = 3$. The product construction yields a sequence of 48 pairs (shown in that section) with the top row 132134 repeated eight times, and the bottom row a slightly scrambled version of 10111000 repeated six times. The next problem is to assign card values to these pairs. This was easy for the binary part, using hearts and diamonds for 1 (red) and clubs and spades for 0 (black). This is a primitive lifting. The lifting problem is harder for other values. We discard kings and think of the other card values as $1, 2, 3, \ldots, 12$. The first step was to lift the sequence 132134 (on an alphabet with four symbols) to six distinct symbols: 143256.

In fact, we can prove that the following lifting procedure always works. Suppose that we have a sequence of digits such that a window of width k gives a distinct relative order as it is run along. Take the highest digit (it may be repeated several times—just choose one) and replace it with $k!$. Take the next highest digit (not counting the $k!$ factorial just created) and replace it by $k! - 1$, and so on. Thus, working right-to-left in 132134 with $k = 3$, we get successively $132136, 132156, 142156, 143156$, and finally, 143256. If we had replaced equal digits working left-to-right, the result would have been $2\,5\,3\,1\,4\,6$. Both final sequences have successive groups of three spanning all possible relative orders.

Now consider two adjacent copies of $1\,4\,3\,2\,5\,6\,1\,4\,3\,2\,5\,6$. Each "1" can be assigned to one of $\{1, 2\}$, each "2" to one of $\{3, 4\}$, and so on, with each "6" assigned to one of $\{11, 12\}$. Choosing the lower possibility first gives

1	4	3	2	5	6	1	4	3	2	5	6
1	7	5	3	9	11	2	8	6	4	10	12.

This pattern is repeated four times, and then the suits are assigned as explained above. We should point out that consecutive values in $\{1, 2\}, \{3, 4\}$, etc. can be interchanged to make things look more random (as we did in Example 1).

An example of a lumping problem appears in adapting an arrangement of the numbers $1, 2, 3, \ldots, k!$ with the property that the relative order of each k-tuple is distinct into an arrangement of values in the alphabet $\{1, 2, 3, \ldots, k+1\}$ into a sequence with the same property. For example, when $k = 3$, $1\,4\,3\,2\,5\,6$ can be lumped to $1\,3\,2\,1\,3\,4$. It is easy to see that maximal-length permutation sequences cannot be lumped to the alphabet $\{1, 2, 3, \ldots, k\}$. For a long time the best that had been proved was $\{1, 2, 3, \ldots, 3k/2\}$ (see [8, 9]). The conjecture that it is always possible with $\{1, 2, 3, \ldots, k+1\}$ was recently solved by Dr. J. Robert Johnson of the Department of Mathematics, Queen Mary College, London. He shows that only $k + 1$ distinct values are required to form a universal cycle of length k.

Lifting and lumping problems arise all over the subject (see [1] for more examples). We would love to see some theory developed for this problem.

Coding and Neat Generation

We have not dealt with one aspect of the applications herein. Given the audience's information, how does the performer know what the cards are? We have treated this at some length in our book on mathematics and magic tricks [2]. However, the performer may have the order of the deck available, coupled to the possible patterns. This availability may be through memory (mnemonics), an assistant, or a hidden list. In our popular talks, we often just say, "The performer has the information written on his sleeve."

As an indication of the methods presently available, we record a novel approach due to our student Gier Helleloid. It allows a neat decoding for any de Bruijn sequence.

Example 7 (Coding a binary de Bruijn cycle). Begin with a fixed de Bruijn cycle of total length m and window width k. We need not have $m = 2^k$. The problem is to assign card values so that the binary color pattern codes the card values in a simple way. Helleloid proposes using a simple *standard* order of the m cards and then using the binary pattern (as a binary number) to determine which card in the standard order goes next.

This is most easily explained by example. Consider the binary de Bruijn cycle 00011101 with $m = 8$ and $k = 3$. Form the standard order of an eight-card deck:

position	0	1	2	3	4	5	6	7
card	AC	2C	AS	2S	AD	2D	AH	2H.

Here the positions have been labeled (from left to right) 0, 1, 2, 3, 4, 5, 6, 7 and AC stands for the ace of clubs, and so on. Helleloid's rule says to rearrange the standard order as

$$AC \quad 2C \quad 2S \quad 2H \quad AH \quad 2D \quad AS \quad AD.$$

Thus, the first window 000 of our de Bruijn sequence says to use the card in position 0 of the standard order (AC). The next window 001 says to use the card in position 1 of the standard order $(2C)$ next. The next window 011 says to use the card in position 3 of the standard order $(2S)$ next, and so on. The scheme works, provided that the standard list has all the black cards first and all the red cards last.

For performance, you must be able to easily determine which card is at position j on the standard list. Thus, if the pattern 101 shows when everyone with a Red card is asked to stand, the performer translates $101 = 5$, and on the standard list, card 5 is $2D$. This is possible (and even easy), provided that the standard list is simple, e.g., for $m = 32$, $k = 5$, we could use 1–8 of clubs, 1–8 of spades, 1–8 of diamonds, and 1–8 of hearts. To continue beyond the first card, a de Bruijn sequence that can easily be "run forward" is essential. The shift register sequences discussed in our references are one simple solution. In this case, the next binary digit is a linear combination of the last few. While we know how to do this for de Bruijn sequences, we do not know of analogous procedures for any of our other constructions. Again, we feel it must be possible. There is a fair amount of worthwhile research to be done here.

Appendix: Proof of Main Theorems, and Somewhat More

In this appendix, we give a proof of Theorem 1 (both windows of equal width) and Theorem 2 (windows of possibly distinct widths). These theorems involve universal *cycles* (going around the corner). They also involve a restriction: one of the cycles must have repeated symbols for all of a window width.

Our proof proceeds by constructing a completely general product (with no restriction on repeated symbols) of two *sequences* (not going around the corner). This is Theorem 3, stated below. Theorems 1 and 2 follow as corollaries.

To state Theorem 3, we need some simple notation. Let

$$\overline{x} = x_1 x_2 \ldots x_R, \quad \overline{y} = y_1 y_2 \ldots y_S, \quad R = rd, \quad S = sd,$$

with d being the greatest common divisor of R and S. Thus, r and s have no common divisor (greater than 1). The symbols x_i, y_j are treated as distinct variables. In the corollaries, they may be set to convenient values (e.g., zero or one).

Construction 1. Construct a two-line array with the top row drawn from the x_i and the second row drawn from the y_j. Both rows will contain RS symbols.

- Top Row. Repeat $x_1 x_2 \ldots x_R$ a total of S times.

- Second Row. Form Y^- by repeating $y_1 y_2 \ldots y_s$ a total of r times and then deleting the final occurrence of y_s. Thus, Y^- has length $rs - 1$. Then, form a sequence of length RS by repeating Y^- a total of d times and adding a total of d repetitions of the symbol y_s at the end.

Example 8. Suppose that $\overline{x} = x_1 x_2$ and $\overline{y} = y_1 y_2 y_3 y_4$. Then, $R = 2$, $S = 4$, $d = 2$, $r = 1$, and $s = 2$. The construction gives an array of total length 8:

$$
\begin{array}{cccccccc}
x_1 & x_2 & x_1 & x_2 & x_1 & x_2 & x_1 & x_2 \\
y_1 & y_2 & y_3 & y_1 & y_2 & y_3 & y_4 & y_4.
\end{array}
$$

Note that each x_1 occurs with each of y_1, y_2, y_3, y_4 exactly once, and this is true for x_2 as well. Theorem 3 says this happens in general.

Theorem 4.3. *Let $\bar{x} = x_1 x_2 \ldots x_R$ and $\bar{y} = y_1 y_2 \ldots y_S$ be strings of distinct symbols. Then, the construction above produces a two-line array of length RS where each pair*

$$\frac{x_u}{y_v}, \ 1 \le u \le R, \ 1 \le v \le S,$$

appears exactly once.

Proof: To check all details, we introduce notation for the blocks of x symbols in the top row and for the blocks of y symbols in the second row. Define

$$X = \overbrace{\bar{x}\,\bar{x} \ldots \bar{x}}^{s}, \quad Y = \overbrace{\bar{y}\,\bar{y} \ldots \bar{y}}^{r}, \quad Y^- = \overbrace{\bar{y}\,\bar{y} \ldots \bar{y}}^{r-1} y_1 y_2 \ldots y_{s-1}.$$

Thus, X has length rsd, as does Y, while Y^- has length $rsd - 1$.

Next, define

$$X_i = X, \quad Y_i^- = Y^-, 1 \le i \le d, \quad Z = \overbrace{y_s y_s \ldots y_s}^{d}.$$

Finally, the array defined by the construction is

$$
\begin{array}{cccccc}
X_1 & X_2 & \ldots & X_{d-2} & X_{d-1} & X_d \\
Y_1^- & Y_2^- & \ldots & Y_{d-2}^- & Y_{d-1}^- & Z.
\end{array}
$$

Note that each row contains RS symbols. We show that each pair

$$\frac{x_u}{y_v}, 1 \le u \le R, 1 \le v \le S,$$

occurs exactly once. There are two cases.

Case 1. The indices of the x_u that are paired with y_v in Y_1^- are $u = v, v + sd, v + 2sd, \ldots, v + isd, \ldots, v + (r-1)sd$ where here, and in what follows, we assume that index addition is done modulo rd, and instead of 0, we use rd. In Y_2, y_v is paired with x_u for $u = v - 1, v - 1 + sd, \ldots, v - 1 + isd, \ldots, v - 1 + (r-1)sd$. In general, in Y_j, y_v is paired with x_u for $u = v - j + 1 + isd, 0 \le i \le r - 1, 1 \le j \le d$. We need to show that all these rd values $v - j + 1 + isd$ are distinct modulo rd.

Suppose that $v - j + 1 + isd \equiv v - j' + 1 + i'sd \pmod{dr}, \quad 0 \le i, i' \le r - 1, 1 \le j, j' \le d$. Thus, $j' - j + (i - i')sd \equiv 0 \pmod{rd}$. This implies that $j' - j \equiv 0 \pmod{d}$ because $\gcd(r, s) = 1$, which in turn implies that $j = j'$. From this we now conclude that $(i - i')sd \equiv 0 \pmod{rd}$. Consequently, we have $(i - i')s \equiv 0 \pmod{r}$, which implies that $i = i'$. Hence, all the rd indices are distinct, so y_v is paired with every possible x_u exactly once, when $v \ne sd$.

Case 2 ($v = sd$). In Y_1^-, y_{sd} is paired with x_u for $u = sd, 2sd, \ldots,$ $(r-1)sd$. In general, in Y_j^-, y_{sd} is paired with x_u for $u = isd - j + 1$, $1 \le i \le r - 1$, $1 \le j \le d$. Also, at the end of the sequence, y_{sd} is paired with the last d symbols of the top row of the array, namely, x_u for $u = dr - d + 1, dr - d + 2, \ldots, dr - 1, dr$.

If $isd - j + 1 \equiv i'sd - j' + 1 \pmod{rd}$, then $j' - j + (i - i')sd \equiv 0$ $\pmod{rd}$. As before, this implies that $j = j'$ and $i = i'$, so all these $(r-1)d$ indices are distinct.

Now suppose that $isd - j + 1 \equiv dr - m \pmod{rd}$, $0 \le m \le d - 1$. Thus, $isd - j + 1 \equiv -m \pmod{rd}$, from which it follows that $j - 1 \equiv m$ $\pmod{d}$, and finally, that $j = m + 1$. Hence, $isd \equiv 0 \pmod{dr}$, which implies that $i \equiv 0 \pmod{r}$, a contradiction.

Consequently, y_{sd} is paired with every possible x_u exactly once. This completes Case 2 and the theorem is proved.

Because Theorem 2 is more general than Theorem 1, we need only prove Theorem 2.

Proof of Theorem 2: Take $\bar{x}$ to be an arbitrary universal cycle. It need not have maximal length. Take $\bar{y}$ to be a universal cycle of window length k. We assume that $\bar{y}$ has a block of k repeated symbols that we take to be 0 for notational simplicity. These appear as the last k symbols of $\bar{y}$. Proceed with the construction as above. What has to be checked is that the following hold:

1. When $y_{sd} = 0$ is removed from the end of Y_i to form Y_i^-, then as the window moves across the boundary between Y_i^- and Y_{i+1}^- in $\ldots Y_i^- Y_{i+1}^- \ldots$ we only lose one copy of the block $\overbrace{000 \ldots 0}^{k}$.

2. Because our construction has $y_{sd} = 0$, we have $Z = \overbrace{000 \ldots 0}^{d}$. Thus, the second row of the array ends with $\overbrace{0000 \ldots 00}^{k-1+d}$. Because $y_1 \neq 0$ (otherwise $\bar{y}$ would have two blocks equal to $\overbrace{000 \ldots 0}^{k}$), as our window of width k goes around the corner in $\ldots Y_{d-1}^- Y_d^- Z Y_1^- \ldots$, we pick up exactly d extra copies of the block $\overbrace{000 \ldots 0}^{k}$.

Therefore, our construction preserves all necessary occurrences of the required k-tuples in the product. This completes the product construction of the universal cycles $\bar{x}$ and $\bar{y}$, and the proof of Theorem 2 is complete. $\qquad\square$

Acknowledgments. Our work on de Bruijn sequences started with card tricks invented with Ronald Wohl. Martin Gardner encouraged us to write these up. These results were presented as a talk at the first Gathering for Gardner (Atlanta, 1993). We thank Geir Helleloid for letting use his nice de Bruijn coding in the "Coding and Neat Generation" section, Hal Fredericksen and Al Hales for comments incorporated in the present version, and Steve Butler for carefully reading the final manuscript. This research was supported in part by NSF grants DMS-0505673 and CCR-0310991.

Bibliography

[1] Fan Chung, Persi Diaconis, and Ron Graham. "Universal Cycles for Combinatorial Structures." *Discrete Mathematics* 110 (1992), 43–59.

[2] Persi Diaconis and Ron Graham. *From Magic to Mathematics—and Back.* To appear.

[3] Persi Diaconis and Ron Graham. Unpublished manuscript.

[4] Hal Fredricksen. "A Survey of Full-Length Nonlinear Shift Register Cycle Algorithms." *SIAM Review* 24 (1982), 195–212.

[5] Martin Gardner. *Fractal Music, Hypercards, and More. . . .* New York: W. H. Freeman and Company, 1992.

[6] Solomon Golomb. *Linear Shift Register Sequences*, Revised Edition. Walnut Creek, CA: Aegean Park Press, 1982.

[7] Stuart Hampshire. *The Game of Tarot.* London: Duckworth Press, 1980.

[8] Glenn Hurlbert. "On Universal Cycles for k-subsets of an n-set." *SIAM Journal on Discrete Mathematics* 7:4 (1994), 598–604.

[9] Glenn Hurlbert and Garth Isaac. "Equivalence Class Universal Cycles for Permutations." *Discrete Mathematics* 149 (1996), 123–129.

[10] Glenn Hurlbert, personal communication.

[11] Brad Jackson. "Universal Cycles of k-subsets and k-permutations." *Discrete Mathematics* 117:1–3 (1993), 141–150.

[12] Brad Jackson, personal communication.

[13] Donald E. Knuth, *The Art of Computer Programming*, Volume 4, Fascicle 2 (Generating All Tuples and Permutations). Reading, MA: Addison-Wesley Professional, 2005.

A Lifetime of Puzzles

[14] Donald E. Knuth, *The Art of Computer Programming*, Volume 4, Fascicle 3 (Generating All Combinations and Partitions). Reading, MA: Addison-Wesley Professional, 2005.

[15] Sherman K. Stein. *Mathematics: The Man-Made Universe*, Third Edition. San Francisco: W. H. Freeman, 1976.

Part II

In Hindsight

Tangram: The World's First Puzzle Craze

Jerry Slocum

The world's first puzzle craze occurred almost 200 years ago, during the years 1817 and 1818, when transportation and communication consisted of sailing ships and horse-drawn carts and carriages. The *Tangram*, a seven-piece, put-together puzzle invented in China between 1796 and 1802, was taken to London on "China Trade" ships and soon became a fashionable puzzle craze in England, Europe, and America. This remarkable event was the world's first puzzle craze. The *Tangram* is still popular and has been used in schools worldwide since the 1860s to help students learn, while having fun.

Among the international celebrities who amused themselves with the *Tangram* are Napoleon Bonaparte (more about him later), Lewis Carroll, Edgar Allan Poe, Hans Christian Anderson, and English scientist Michael Faraday, not to mention our good friend Martin Gardner [1, Chapters 3 and 4].

The Tangram

The *Tangram*, a two-dimensional rearrangement puzzle, is formed by dissecting (cutting) a square into seven pieces, called *tans*. (See

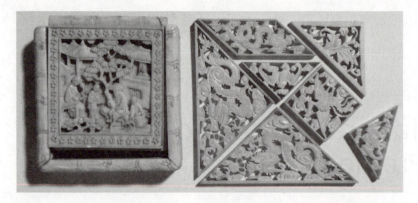

Figure 1. An ivory *Tangram* box and seven puzzle pieces made in China for export.

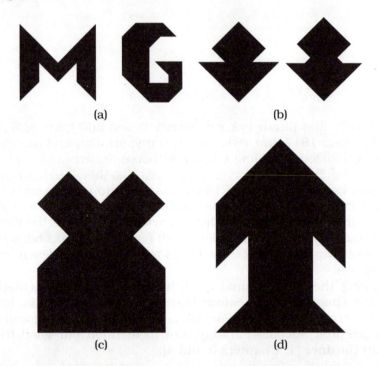

(a)

(b)

(c)

(d)

Figure 2. Some *Tangram* challenges for the reader. Solutions are at the end of the article. (a) Martin Gardner puzzles: make each initial individually. (b) *Tangram* paradox. (c) A difficult Chinese *Tangram* puzzle by Ch'lu Llang-pal from 1885. (d) A difficult Russian *Tangram* puzzle by V. I. Obreimov from 1884.

Figure 1.) The pieces can be rearranged to form thousands of different figures of people in motion, animals, letters of the alphabet, geometric shapes, and the universe. The puzzle is to assemble all seven pieces, without overlap, to form a given problem figure (such as the ones in Figure 2).

Several rearrangement puzzles were invented before the *Tangram*, and many have been invented since, but the *Tangram* has turned out to be the most popular by far. The seven pieces are simple shapes: two small triangles, one medium-sized triangle, two large triangles, a rhomboid, and a square. It is unique among rearrangement puzzles in its ability to transform these simple geometric pieces into charming, elegant, sophisticated, and sometimes paradoxical figures. The silhouette problems are presented in books or on cards that accompany the *Tangram*. Or you can create your own designs, limited only by your imagination. The inscrutable face of a Chinese emperor, the elegance of a bird in flight, and puzzling paradox figures can all be made from the amazing *Tangram*. The puzzle's very simplicity proves most maddening: how can seven simple tans create such extraordinary images and puzzling challenges? In Chinese, the *Tangram* is known as *Ch'i ch'iao t'u*, which translates to "seven ingenious plans" or "picture using seven clever pieces."

Invention of the Tangram

According to Chinese reference literature, Yang-cho-chü-shih (Dim-witted recluse) invented the *Tangram* during the reign of Chia-ch'ing (1796–1820). Recently, however, a second edition of his book, *Tseng-ting ch'i-ch'iao t'u*, dated 1817, has been found, and Yang-cho-chü-shih references Sang-hsia'k'o's book as his source. So, the inventor of the *Tangram* is still unknown. The earliest found example of a *Tangram* is in a silk-covered cardboard box with a handwritten inscription dated April 4, 1802. (See Figure 3.) The box contains a carved ivory *Tangram* that was given to Francis Waln, the third child of Robert and Phebe Waln. Robert Waln was a major ship owner and importer in Philadelphia, with a financial interest in at least twelve ships, trading with Canton, China.

Sang-hsia-k'o (a pen name, meaning "guest under the mulberry tree") compiled the problem figures for the second *Tangram* book, entitled *Ch'i ch'iao t'u ho pi* (*Harmoniously Combined Book of Tangram Problems*) and wrote a preface for it. The preface and the 334 problem figures in the book, published in 1813, were widely

Figure 3. This is the earliest known *Tangram* (left) and its silk-covered case (right). It was given to Francis Waln in 1802.

reprinted in numerous editions by several Chinese publishers for over 100 years. No copies of the 1813 edition have been found in China, Japan, England, Europe, or the USA. However, a replica of the original 1813 Chinese book, including the cover, the text, and 130 of the problems, was discovered in Japan. (See Figure 4.) The copy, with the text in blue and the figures in red, was published in Japan in 1839.

The preface by Sang-hsia-k'o tells a bit about the history of the *Tangram* and the problem figures:

"Its origin lies within the Pythagorean Theorem. Last year Hsü Shu-t'ang traced 160 *Tangram* designs and published them. Mr. Wang I-yüan brought a copy of Hsü Shu-t'ang's booklet and added designs by his younger brother, Ch'un-sheng, to it. The manuscript included about 200 designs. I invented another 100 new designs, which were added to the copy. I didn't want to keep it for my pleasure only, so I decided to publish it for the entertainment of those who also love this game."

Our investigation of the Pythagorean Theorem in Chinese mathematics found no evidence that the *Tangram* was invented or known

Figure 4. A replica, published in Japan, of Sang-hsia-k'o's 1813 Chinese *Tangram* book: title page (left) and sample problem (right).

by ancient Chinese mathematicians. However their method of dissecting a figure and rearranging the pieces to form a new figure was an integral part of Chinese mathematics in the third century A.D. and was the approach used by the Chinese to prove the Pythagorean Theorem. This is one of the roots in the Chinese culture that may well have contributed to the invention of the *Tangram* many centuries later.

The publisher of Sang-hsia-k'o's 1813 book published a new edition in 1815, containing the same preface and the same problems as the earlier edition. A book of solutions was also published, and the books were sold as a pair. The books were made of accordion-folded rice paper sewn together with a string binding. Copies of the 1815 edition were widely distributed, not only in China, but also in England, Europe, and America, and they were responsible for spreading the *Tangram* craze to the Western world. (See Figure 5.)

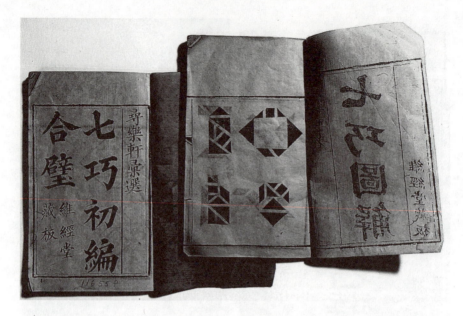

Figure 5. The oldest surviving pair of Chinese *Tangram* problem and solution books was published in 1815.

The Tangram Craze Hits Europe and America

Puzzles made in China of ivory and wood, as well as copies of Sang-hsia-k'o's *Tangram* books, were brought to England and Europe on sailing ships. After the Chinese books reached England, the problems were copied and published and the puzzle quickly became fashionable in London. Its popularity rapidly spread to other European countries. As the German author C. L. A. Kunze describes, "this game, soon after its appearance, had become a favorite amusement in educated families of Northern Germany. The examples came from England and were offered by Hamburg art dealers and, according to information passed by word of mouth, they were very elegant: the figures printed on natural paper with the beautiful Chinese cinnabar (brilliant red), the seven pieces decoratively carved from foreign wood or ivory or mother of pearl, the whole enclosed in cases being lacquered in black and gold."

The credit for making the puzzle fashionable and popular, not only in London, but throughout most of Europe, must be given to a pair of elegant British books: *The Fashionable Chinese Puzzle* and its companion, *Key*, published by John and Edward Wallis and John Wallis, Junior, in March of 1817. (See Figure 6.)

Figure 6. *The Fashionable Chinese Puzzle*, published in London during March 1817 by John Wallis, was extensively copied and spread the *Tangram* craze to Europe and America: cover (left) and sample spread (right).

The problem book includes a hand-colored illustration of a Chinese scene on the cover and a poem called "Stanzas" as the preface. The hand-colored problem drawings looked accurate, and the companion solution book was easy to use. The books used high-quality paper. The poem mentioned, among other things, that the Chinese Puzzle was "the favourite amusement of Ex-Emperor Napoleon."

Napoleon's Tangram Found

One member of the team of researchers supporting the author in the investigation into the history of the *Tangram*, Dic Sonneveld, searched the Internet and libraries and read dozens of books about Napoleon for clues to the mystery of almost 200 years of whether, as Wallis said, the *Tangram* was Napoleon's favorite pastime. On one trip to the Bibliothèque Nationale of France, in Paris, he also visited museums containing Napoleon artifacts in

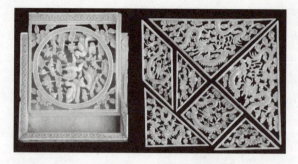

Figure 7. This ivory *Tangram* and a pair of Chinese problem and solution books, dated 1815, once belonged to Napoleon.

and around the city. In one of the museums, he found a display case with Napoleon's beautifully carved ivory *Tangram* case and seven tans (Figure 7), as well as a pair of Chinese books from 1815 containing problems and solutions.

On the Continent

The first *Tangram* book on the Continent was published in France by Grossin of Paris on July 19, 1817, and bore the title *Énigmes Chinoises* (Figure 8). It was copied from Wallis' *Fashionable Chinese Puzzle*. Likewise, the first *Tangram* books published in Switzerland, Italy, the Netherlands, and Denmark all copied Wallis. And while the first *Tangram* books published in America copied the Chinese book, this changed when the books by Wallis arrived. In 1817 most countries had laws forbidding the copying of books inside the country itself, but there were no treaties between countries prohibiting the copying of books published in another country.

In France, in 1817 and 1818, artists improved on the plain outlines used for problems in the Chinese and British books and created beautiful hand-colored problem figures of people and animals that looked like miniature pictures. Cards with the problem drawings were included in boxed sets, along with the seven pieces of the puzzle. (See, for example, Figure 9.)

The *Tangram* craze peaked in France during the first quarter of 1818, as demonstrated by the number of *Tangram* sets and books published, as well as by two elaborate caricatures showing the excesses brought on in France by the *Tangram* craze. The first, *Le Goût du jour No. 45; Le Casse-tête Chinois* [Caricatures Parisiennes], was published in Paris on January 10, 1818, by Chez Mar-

A Lifetime of Puzzles

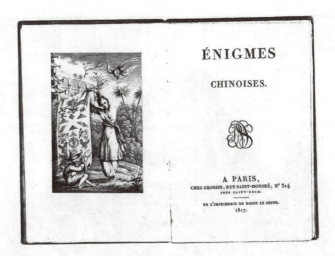

Figure 8. Énigmes Chinoises was the first *Tangram* book published on the European continent.

Figure 9. The portrait of King Henry IV of France was published in Paris in 1818 as one of a set of 16 *Tangram* problem cards.

tinet and shows a couple ignoring their crying baby's needs and the lack of heat in the house while they are staying up all night to solve *Tangram* problems (Figure 10). The Martinet bookshop was famous for its caricatures, and they always had some in the windows, with new ones every fifteen days. The *Tangram* craze ended in France by the end of 1818.

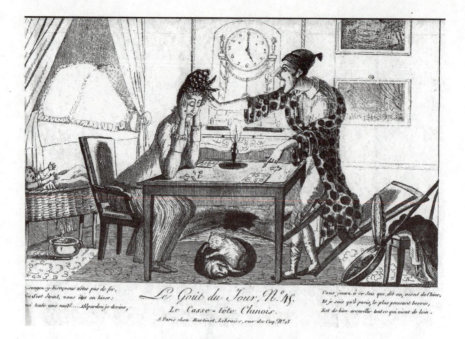

Figure 10. Caricature of life in Paris, published at the height of the *Tangram* craze in January 1818. The caption translates: "Take care of yourself, you're not made of steel. The fire has almost gone out and it is winter. It kept me busy all night. Excuse me, I will explain it to you. You play this game, which is said to hail from China. And I tell you that what Paris needs most right now is to welcome that which comes from far away."

The first Italian *Tangram* book, published in 1817, was a copy of the British book by Wallis. But in 1818, G. Landi of Florence, Italy, produced *Metamorfosi Del Giuoco Detto L'Enimma Chinese* (Metamorphosis of the Game also Known as Chinese Enigma), a beautiful book of 100 miniature pictures of architectural features such as monuments, buildings, fountains, and bridges, that were so artistically made that each problem was a beautiful picture. (See Figure 11.)

Germany became fascinated by the *Tangram* at least six months later than France, with numerous publications during 1818, including two beautiful hand-colored sets of picture problem cards. Although the *Tangram* did not reach the level of craze seen in France, it was much more sustained, with German *Tangram* publications occurring every few years through the end of the nineteenth century. (See Figure 12 for an example.)

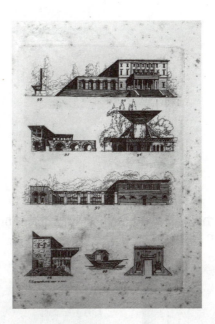

Figure 11. One of ten plates of engraved *Tangram* architecture problems in *Metamorfosi Del Giuoco Detto L'Enimma Chinese*, published in Florence, Italy, in 1818.

Figure 12. These figures are part of 24 problem cards published in Germany with the title *Hieroglyphen oder Bilderschrift* (Game of Mystical Characters).

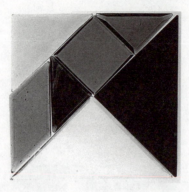

Figure 13. A glass *Tangram*, with a sheet of problem figures, was made in China, but found in Denmark.

Figure 14. *Stort Chinesisht Gätspel: Grandes Énigmes Chinoises* (Large Chinese Puzzle Game) was published in Sweden by Fehr and Muller between 1817 and 1820.

Denmark had a remarkable interest in the *Tangram* during 1818, with four publications. One of the books stated, "Many thousands of sets are sold at different shops in Copenhagen made from mother-of-pearl, ebony, mahogany, and other types of wood, even of glass." See Figure 13 for an example of a glass *Tangram*.

Sweden was also interested in the *Tangram* as the craze swept Europe. A beautiful set of 36 hand-colored problem pictures was published in Sweden. (See Figure 14.)

Across the Atlantic

In America, a pair of Sang-hsia-k'o's 1815 *Tangram* books was given to Captain Edward M. Donnaldson on October 30, 1815, while he was docked in Canton. He brought them to the United States on his ship, *Trader*, and arrived in Philadelphia in February 1816.

Although two books with *Tangram* problems, copied directly from the 1815 *Sang hsia k'o* book, were published in the United States in 1817, there was not nearly as much excitement about the puzzle in America as there was in China, England, and Europe.

Figure 15. One of three *Tangram* sets published by McLoughlin Brothers during the 1870s. The sets also included sixty hand-colored picture problems copied from the French.

The first book, *Chinese Philosophical and Mathematical Trangram*, was published by James Coxe in August 1817; later the same year, a New York publisher, A.T. Goodrich, published a pair of problem and solution books entitled *The New and Fashionable Chinese Puzzle*.

The poor quality of these first *Tangram* books was probably a major factor in the lack of enthusiasm for the puzzle in America. Also, in 1818 the only new *Tangram* book published in America was a copy of Wallis' *Fashionable Chinese Puzzle* by A.T. Goodrich. This book, and the puzzles themselves, continued to be advertised in New York and Boston through the end of 1822. Interest in the puzzles increased during the period from 1865 to 1880, when numerous boxed sets of *Tangram*s were produced by several companies. This increased activity may have been caused by the use of *Tangram*s in schools, which began during the same period. McLoughlin Brothers published a beautiful boxed set with hand-colored problem cards copied from a French edition. (See Figure 15.)

Figure 16. Sam Loyd's *The 8th Book of Tan*, published in New York in 1903, popularized the name "Tangram," as well as the puzzle itself, and contained over 650 problems, 430 of which were invented by Loyd. It also included an imaginative but bogus history of the puzzle that still persists today.

America's greatest puzzle designer, Sam Loyd (1841–1911), designed and published many *Tangram* problems. His first booklet of original *Tangram* problems appeared in 1875, and his famous *Eighth Book of Tan* (Figure 16), with untrue but imaginative stories of the history of *Tangram* and hundreds of original problems, was published in 1903.

Merchandising of the Tangram

According to Carl Crossman's book, *The China Trade*, "Ivory puzzles intrigued every merchant who went to China. These seemingly simple products of clever design and good craftsmanship were made in all shapes and forms and were often described in great detail by Westerners who had purchased them. The puzzles could be bought singly or in groups, either in fabric covered pasteboard boxes or very handsomely decorated lacquer boxes."

The popularity of the *Tangram* in China inspired merchants there to produce plain wooden and ivory *Tangram*s for domestic use (see, for example, Figure 17), and fancy puzzles for export from materials such as ivory, mother-of-pearl, tortoise shell, ebony, mahogany, copper, and even glass.

Pairs of beautiful *Tangram* problem and solution books were sold with intricately carved ivory and mother-of-pearl covers; some books even had the pages covered with hand-painted silk. (See, for example, Figure 18.) Sets of dishes in the form of the seven tans using cloisonné over bronze were also marketed (Figure 19). Minia-

Figure 17. Uncarved ivory *Tangram* pieces in a handy wooden case for use in China.

Figure 18. *Tangram* problem book with ivory cover (left) and silk pages (right) that contain over 340 problem figures.

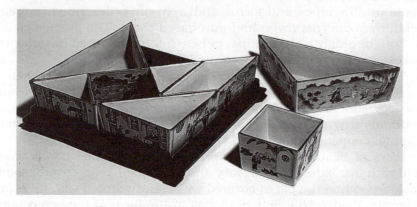

Figure 19. The sides of these ceramic *Tangram* dishes are decorated with colorful scenes of ancient Chinese legends.

ture sets of sandalwood and rosewood *Tangram* tables were produced for export, and beautiful full-sized *Tangram* tables of ironwood with burl inlay (Figure 20) were made to sell in the country itself.

Conclusion

The Chinese have invented many new mechanical puzzles, from the "impossible" Magic Mirror, to puzzle vessels and wire puzzles, but the *Tangram* is the only Chinese puzzle to sweep the western world, become a puzzle craze, and continue to be very popular in schools and with the general public for almost 200 years.

Figure 20. The author in his Slocum Puzzle Museum arranging *Tangram* tables of ironwood with burl inlay, made in China around 1840.

Acknowledgments. This article is a summary of one part of *The Tangram Book* (Sterling, 2003) by Jerry Slocum, with Jack Botermans, Dieter Gebhardt, Monica Ma, Xiaohe Ma, Harold Raizer, Dic Sonneveld, and Carla van Splunteren. The interested reader can find 192 full-color pages of a comprehensive history of the Tangram and more than 2,000 Tangram problems to solve in that book.

Bibliography

[1] Martin Gardner. *Time Travel and Other Mathematical Bewilderments.* New York: W. H. Freemann, 1998.

Solutions

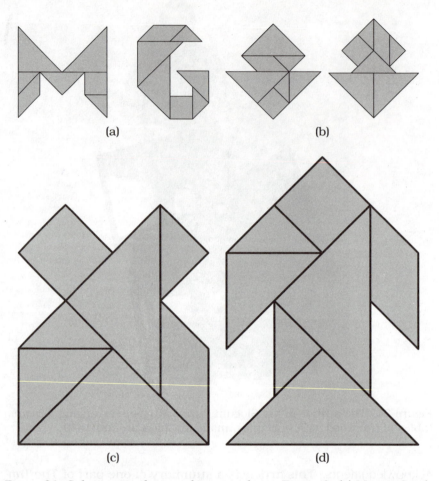

(a)

(b)

(c)

(d)

Figure 21. Solutions to the matching puzzles in Figure 2: (a) Martin Gardner puzzles. (b) *Tangram* paradox. (c) A difficult Chinese *Tangram* puzzle. (d) A difficult Russian *Tangram* puzzle.

A Lifetime of Puzzles

De Viribus Quantitatis by Luca Pacioli: The First Recreational Mathematics Book

David Singmaster

Luca Pacioli (ca. 1445–1517) was born and probably died in (Borgo) San Sepolcro, a small city in southeastern Tuscany (see Figure 1), so he is sometimes called Luca del Borgo. Vasari asserts that he was a student of Piero della Francesca (ca. 1416–1492), also of San Sepolcro, but there is no supporting evidence for this. He was a Franciscan friar from ca. 1475. He was the most famous mathematician of his day, being a leading expositor of the new theory of perspective and the author of the most important mathematical work after Fibonacci.

He taught in San Sepolcro, Venice, Perugia (first professor of mathematics there), Rome, Zara (on the Dalmatian coast, then part of Venice), Naples, Milan, Florence, Pisa, and Bologna. He must also have been in Urbino several times. His life was remarkably

---

This article describes Singmaster's part of a joint presentation with Vanni Bossi at the 6th Gathering for Gardner, 2004. Bossi's part follows immediately after this article.

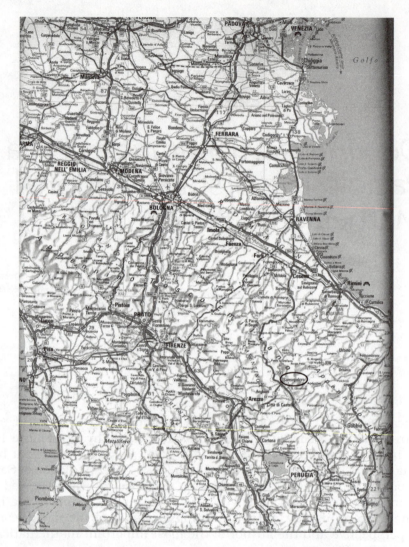

Figure 1. Map of the area where Luca Pacioli lived.

peripatetic, even for the day. There are indications that his superiors in the Franciscan Order advised against his teaching boys, and I wonder if this behavior may have led to his having to move frequently. In Rome, about 1470, he stayed with Alberti, the author of the first book on perspective, and became known as the leading authority on perspective.

A Lifetime of Puzzles

Figure 2. Portrait of Pacioli, attributed to Jacopo de' Barbari.

Pacioli is the earliest mathematician of whom we have a genuine portrait. This splendid picture, shown in Figure 2, is in the Museo Nazionale Capodimonte in Naples, apparently by Jacopo de' Barbari, probably done in Venice, about the time of publication of Pacioli's *Summa* of 1494 (or of *De Divina Proportione* in 1509). The books depicted are his *Summa* and the first printed Euclid of 1482. It has been claimed that the youth on the side is Albrecht Dürer, possibly a self portrait, and that the glass rhombicuboctahedron was done by Leonardo. A biographer of Pacioli objects to this theory because the painting is done with great accuracy and the young man has blue eyes, which Dürer did not. But we know Dürer studied with de' Barbari, and it is conjectured that Dürer studied perspective with Pacioli, possibly in 1506.

It is claimed that Pacioli is the second figure from the right in Piero della Francesca's *Madonna and Child Enthroned with Saints and Angels*, done about 1472 (Figure 3). The depicted saint is St. Peter Martyr, distinguished by the gash on his head.

Figure 3. *Madonna and Child Enthroned with Angels and Saints* by Piero della Francesca (Brera Altarpiece).

In 1494, Pacioli published the greatest mathematical work since Fibonacci (1202): *Summa de Arithmetica, Geometria, Proportioni et Proportionalità* (Venice, 1494). This is a massive book, 616 large pages, too large for my scanner! (See sample pages in Figures 4 and 5.) Part II, ff. 68v–73v, prob. 1–56, is essentially identical to Piero della Francesca's *Trattato*, ff. 105r–120r. This was the first printing of many mathematical concepts, e.g., algebra, double-entry bookkeeping, and pictures of some Archimedean polyhedra. He asserted that cubics and quartics cannot be solved by the methods used for quadratics, which inspired the development of algebra. Pacioli spent some time in Venice supervising the publication.

Sūma de Arithmetica Geo-
metria Proportioni τ Pro-
portionalita.

Continentia de tutta lopera.

De numeri e misure in tutti modi occurrenti.
Proportioni e pportiōalita anotitia del.5? de Eucli
de e de tutti li altri soi libri.
Chiaui ouero euidentie numero.13.p le q̃tita continue pportiōali del.6?e.7? de Euclide extratte.
Tutte le pti del algorismo: cioe releuare . ptir. multi-
plicar.sūmare.e sotrare cō tutte sue pue i sani e rot-
ti.τ radici e progressioni.
De la regula mercantesca ditta del.3.e soi fōdamen-
ti con casi exemplari per cēm? 8.G.guadagni: perdi
te: transportationi: e inuestite.
Partir.multiplicar.summar.e sotrar de le proportio
ni e de tutte sorti radici.
De le.3.regole del cata yn ditta positiōe e sua origie.
Euidentie generali ouer conclusioni n?66.absoluere
ogni caso che per regole.ordinarie nō si podesse.
Tutte sorte binomij e recisi e altre linee irratiōali del
decimo de Euclide.
Tutte regole de algebra ditte de la cosa e lor fabri-
che e fondamenti.
Compagnie i tutti modi.e lor partire.
Socide de bestiami. e lor partire
Fitti: pesciōi: cottimi: liuelli: logagioni: e godimenti.
Baratti i tutti modi semplici: composti: e col tempo.
Cambi reali.secchi.fittitij.e diminuti ouer comuni.
Meriti semplici e a capo danno e altri termini.
Resti.saldi.sconti.de tempo e denari e de recare a un
di piu partire·
Or.argēti.e lloro affinare. e carattare
Molti casi e ragioni straordinarie varie e diuerse a
tutte occurentie commo nella sequente tauola ap-
pare ordinatamente de tutte.
Ordine a saper tener ogni cōto e scripture e del qua
derno in vinegia.
Tariffa de tutte vsange e costumi mercanteschi in tut
to el mondo.
Pratica e theorica de geometria e de li.5.corpi regu-
lari e altri dependenti.
E molte altre cose d grandissimi piaceri e frutto cō-
mo difusamente per la sequente tauola appare.

Figure 4. First page of the contents of the *Summa*.

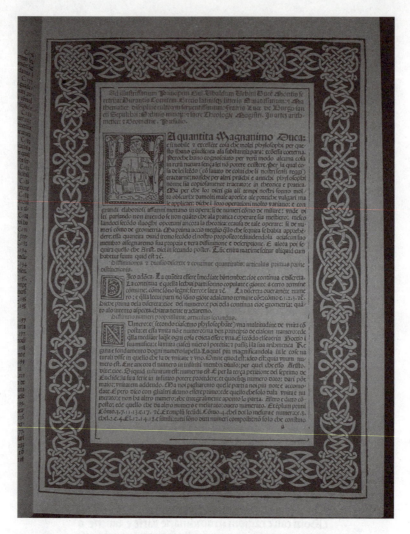

Figure 5. Title page of the *Summa*.

Novel recreational topics in the *Summa* are the following:

- First printed version of the Problem of Points. (If you agree to play until one player has won three times, how do you divide the stakes if you have to stop, say, when the score is 2–1? This later was one of the sources of probability theory.)

- First Locate a Well Equidistant from the Tops of Three Towers problem.

A Lifetime of Puzzles

- First printed pictures of Archimedean polyhedra, namely, the truncated tetrahedron and cuboctahedron. He also mentions the icosidodecahedron and the truncated icosahedron.

Pacioli was a professor at Milan in 1496–1499. This was the high point of his career, being a leading member of the glittering intellectual court of Lodovico Sforza. He was a good friend of Leonardo da Vinci (1452–1519); they even lived together! Pacioli is our leading witness to Leonardo's work at this time, particularly the *Last Supper* during 1495–1497, and he probably advised on the perspective of the painting. Certainly Pacioli stimulated Leonardo's interest in perspective, and it is possible that Leonardo's famous drawing of the proportions of the human body (Figure 6) was inspired by Pacioli's comment on classical architecture; "For in the human body they found the two main figures . . . , namely the perfect circle and the square." Pacioli wrote his *De Divina Proportione* here in 1498, and Leonardo drew splendid pictures for it, though it was not published (in an expanded form) until 1509. (See Figures 7–9.)

Pacioli seems to have made several sets of models of the polyhedra in his book, though we don't know whether Leonardo assisted in making them or used them for his drawings. Pacioli also wrote much of *De Viribus Quantitatis* in Milan.

The printed version of *De Divina Proportione* (see Figures 10 and 11) included a version of Piero della Francesca's *Libellus de Quinque Corporibus Regularibus* of ca. 1487 and the handsome and often reproduced geometric designs for letters of the alphabet.

When the Sforzas were overthrown by the French invasion in 1499, Pacioli and da Vinci moved to Florence, originally lodging in the same house. Pacioli taught at the Universities of Florence and Pisa during 1499–1507, but he may have taken some time out to teach at Bologna, possibly during June–July 1501, and possibly going there to meet Dürer in 1506.

In 1508–1509, Pacioli returned to Venice to publish his *De Divina Proportione* and gave a renowned lecture on the Fifth Book of Euclid. Erasmus was also in Venice at the time and may have attended Pacioli's lecture; he certainly satirizes Pacioli and his works in *In Praise of Folly*. Pacioli may have met Dürer, who was in Venice in 1505–1507. Dürer seems to have taken ideas from Piero della Francesca's *De Prospettive Pingendi*, which was at Urbino, and Pacioli is the most likely person to have shown it to Dürer.

For more details of Pacioli's life and work, see the Appendices.

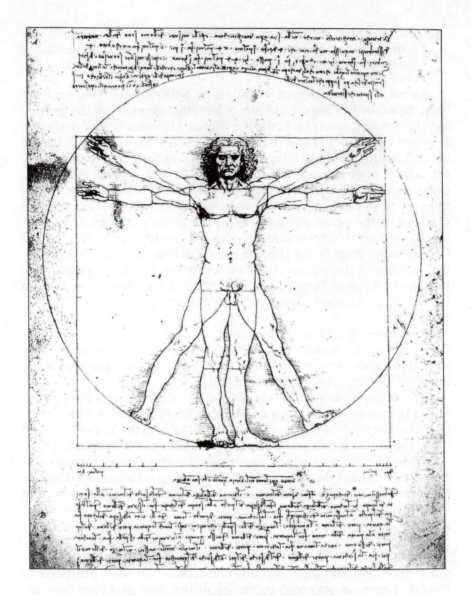

Figure 6. Leonardo's geometric man.

A Lifetime of Puzzles

SSendo Ex̅ .D adi
viiii.de febraro de
noſtra ſalute ꝗ̃ani.
1 4 9 8. correndo
nelinſpugnabile ar
ce de lindyra no
ſtra Citta de Mila
no digniſſimo luo ·
go de ſua ſolita reſidentia ala preſentia di ꝗ̃lla
conſtituto in lo laudabile e ſcientifico duello o'
da molti de ogni grado celeberrimi e ſapientiſſi
mi accompagnata : ſi religioſi como ſeculari o'
deliquali : aſſiduc la ſua magnifica corte ha
bunda : del cui numero oltre le R̅ᵐᵉ⁺. S. di ueſco
ui prothonotarii e abbati fuoron del noſtro
ſacro ſeraphico ordine el R̅ᵈᵒ padre e ſublime
theologo maeſtro Gometto col digniſſimo de

Figure 7. Title page of the *De Divina Proportione* manuscript.

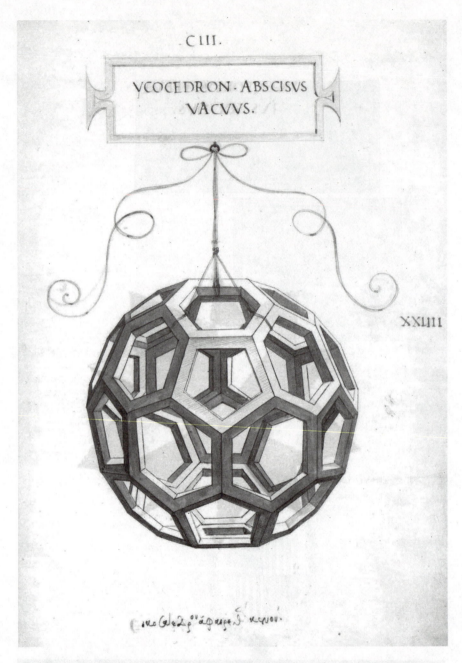

Figure 8. Picture of the truncated icosahedron, from the *De Divina Proportione* manuscript.

A Lifetime of Puzzles

Figure 9. Several pictures from the *De Divina Proportione* manuscript, from a poster.

Díuína

proportione

O pera a tutti glingegni perſpi
caci e curioſi neceſſaria Que cia
ſcun ſtudioſo vi Philoſophia:
Proſpectiua Pictura & ſculptu
ra: Architectura: Muſica: e
altre Mathematice: ſua
uiſſima: ſottile: e ad
mirabile doctrina
conſequira: e de
lectaraſſi:cóva
rie queſtione
de ſecretiſſi
ma ſcien
tia.

M. Antonio Capella eruditiſſ. recenſente:
A. Paganius Paganinus Characteri
bus elegantiſſimis accuratiſsi
me imprimebat.

Figure 10. Title page of the printed version of *De Divina Proportione*.

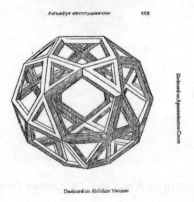

Figure 11. Picture of the truncated icosahedron, from the printed version
of *De Divina Proportione*.

The De Viribus Quantitatis

This article is primarily about the *De Viribus Quantitatis* of ca. 1500. This is an Italian manuscript in Codex 250, Biblioteca Universitaria di Bologna. Pacioli petitioned for a privilege to print this book in 1508, and a problem has a date of 1509, but he seems to have been working on the manuscript since 1496. The title is a bit cryptic, but I think the best English version is *On the Powers of Numbers.*

Dario Uri has photographed the entire manuscript and enhanced the images and put them all on a CD. This CD has 614 images, including the insides of the covers. The photographs are often more legible than the microfilm version (compare Figures 12 and 13), but the folio numbers are often faint, sometimes illegible. He has put some material up on his website,[1] which includes the indexes and a number of the most interesting items, with his comments and diagrams of later examples of the puzzles. All figures in this article from *De Viribus Quantitatis* are from Uri's photos, unless otherwise specified.

There is a transcription by Maria Garlaschi Peirani, with a preface and editing by Augusto Marinoni (Ente Raccolta Vinciana, Milano, 1997).[2] I will cite this text as *Peirani*. The transcription is not exactly literal, in that Peirani has expanded abbreviations and inserted punctuation, etc. Also, Peirani seems to have worked from the microfilm or a poor copy, as she sometimes says the manuscript has an incorrect form that she corrects, but Dario Uri's photo clearly shows that the manuscript has the correct form. Peirani uses the problem numbers and names in the manuscript (see comment below about these differing from those in the index), but with some amendments. I give problem names as in the manuscript, with some of Peirani's amendments.

This is the first large work devoted to recreational mathematics. There are three parts. It opens with a table of contents, which turns out not to be very dependable. The actual text, with folios numbered with Arabic numerals, opens with some fragments of dedication, possibly leaving room for some fancy initial lettering. Part 1 is 81 arithmetical recreations. Part 2, "*della virtù et forza geometrica*" (On Geometric Virtue and Power), contains 134 geometrical and topological problems. Part 3 contains several hundred

---

[1] http://digilander.libero.it/maior2000/

[2] Dario Uri says it can be bought from Libreria Pecorini, 48 foro Buonaparte, Milano; tel: 02 8646 0660; fax: 02 7200 1462; web: http://www.pecorini.com.

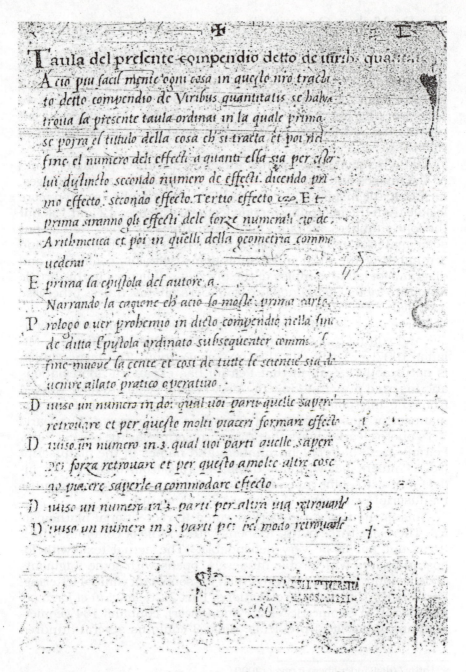

Taula del presente compendio detto de viribus quan...

Acio piu facil mente ogni cosa in quello nro tracta
to detto compendio de Viribus quantitatis se habia
troua la presente taula ordinai in la quale prima
se pojra el tittulo della cosa ch si tracta et poi nel
fine el numero deli effecti a quanti ella sia per effer
lui dylincto secondo numero de effecti. dicendo pri
mo effecto. secondo effecto. Tertio effecto cro. Et
prima siranno gli effecti dele forze numerali cro de
Arithmetica et poi in quelli della geometria comme
uedenui

E prima la epistola del autore a
  Narrando la cagione ch acio lo mosse prima carte

P rologo o uer prohemio in dicto compendio nella fine
  de ditta Epystola ordinato subsequenter comm.
  fine mueue la cente et cosi de tutte le scientie sia de
  uenire allato pratico operattuo

D iuiso un numero in do: qual uoi parti quelle saper
  retrouare et per queste molti piaceri formare effecti

D iuiso un numero in 3. qual uoi parti quelle saper
  per forza retrouare et per questo amolte altre cose
  ao piacere saperle acommodare effecto

D iuiso un numero in 3. parti per altra uia retrouare   3

D iuiso un numero in 3. parti per uel modo retrouare   4

Figure 12. First page of the index, from the microfilm: F. Ir = Uri 4 = Peirani 3.

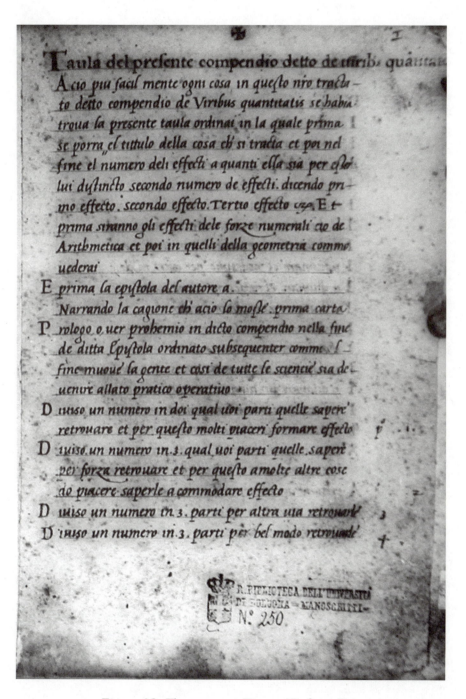

Figure 13. The same as Figure 12, from Uri.

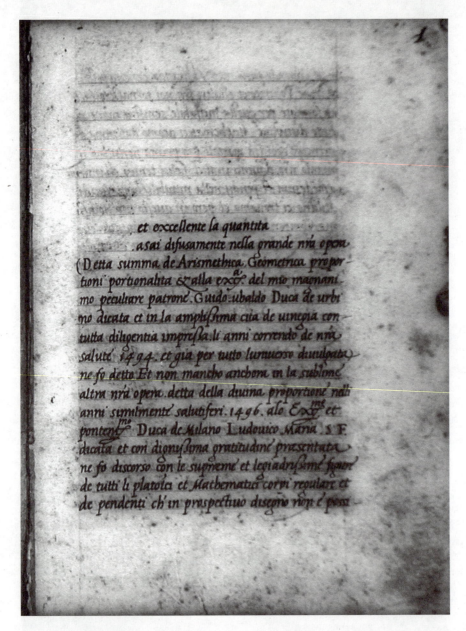

... , et excellente la quantita ...
... asai difusamente nella grande nra opra
(Detta summa de Arismethica. Geometrica proportioni portionalita et alla ex̄c̄o. del mio magnani-
mo peculiare patrone. Guido ubaldo Duca de urbi
no dicata et in la amplissima cita de uinegia con
tutta diligentia impressla. li anni correndo de nra
salute 1494. et gia per tutto luniuerso diuulgata
ne fo detto Et non mancho anchora in la sublime
altra nra opera detta della diuina proportione nelli
anni similmente salutiferi. 1496. alo Ex̄m̄o et
pontem̄o Duca de Milano Ludouico Maria S.F.
dicata et con dignissima gratitudine presentata
ne fo discorso con le supreme et legiadrissime signor
de tutti li platoici et Mathematici corpi regulare et
de pendenti ch' in prospectiuo disegno non e possi

Figure 14. Title page: F. 1r = Uri 30 = Peirani 21.

A Lifetime of Puzzles

proverbs, poems, riddles, and tricks (i.e., physical recreations, conjuring, etc.), in several sections:

- Very Useful Moral Items like Proverbs (23 rhyming couplets);

- Lament of a Lover for a Maid (27 rhyming couplets, based on the letters of the alphabet, and some extra couplets);

- Very Useful Mercantile Items and Proverbs (83 items);

- On Literary Problems and Enigmas (about 80 items);

- Common Problems to Exercise the Ingenuity and for Relaxation (222 items).

Part 1, "*Delle forze numerali cioe di Arithmetica*," is described in A. Agostini, "*Il 'De viribus quantitatis' di Luca Pacioli*," *Periodico di Matematiche* 4:4 (1924), pp. 165–192 (also separately published with pp. 1–28). Agostini's descriptions are sometimes quite brief: unless one knows the problem already, it is often difficult to figure out what is intended. Further, he sometimes gives only one case from Pacioli, while Pacioli does the general situation and all the cases. There are 81 problems in part 1, but the index lists 120!

I had copied Part 1 from the microfilm at the Warburg Insitute and had seen that there were some other interesting problems in Part 2, especially one diagram of a topological puzzle, but I only copied a few pages. I found it difficult to read the Italian (many words are run together and/or archaic), and referenced diagrams are lacking. When I did work on the topological problem, I saw that the following pages would be interesting, but the Warburg had mislaid the microfilm. With the later transcription, I could only read about two further puzzles of this sort. Dario Uri was able to carry on, and found the Chinese Rings and about a dozen other examples of the earliest known topological puzzles. Quite a number of problems, some clearly of interest, remain obscure.

Interesting Recreational Material in De Viribus Quantitatis

In the following, I give the folio numbers from the manuscript, the image number in Uri's photos, and the pages in Peirani's transcription.

The manuscript includes the first European mention of the Blind Abbess and Her Nuns: Ff. IVv–Vr = Uri 12 = Peirani 8. The Index has (Part 1) Problem 89:

De uno abate ch' tolse aguardar certo monasterio de monache in levante contandole sera e matina per ogni verso tante et pur daloro schernito desperato la bandona (Of an abbot who tries to guard a certain monastery of monks in the Levant by counting evening and morning the same on each side and how the sneering desperados abandoned it).

This is the problem where there is a 3×3 square, with three nuns in each exterior cell, and the Blind Abbess can only count the number of nuns along each side, namely nine. But this allows considerable variation in the number of nuns present. At least one Arabic version is known.

In the *Summa*, ff. 97r–97v, no. 34, Pacioli gives a general discussion of the use of 1, 3, 9, 27, 81, 243, ... as weights. In *De Viribus Quantitatis*, f. XIIIv = Uri 29 = Peirani 20, the Index for the third part has Problem 85, *De far 4 pesi che pesi fin 40* (To make four weights which weigh to 40), but at the end, Pacioli says this problem is in "*libro nostro*," i.e., the *Summa*: Cf. Agostini, p. 6.

In the *Summa*, ff. 97v–98r, no. 35, Pacioli gives the problems of using five cups to pay daily rent for 30 days. It uses cups of weights 1, 2, 4, 8, and 15. In *De Viribus Quantitatis*, f. XIIIv = Uri 29 = Peirani 20, the Index for the third part has Problem 86, *De 5 tazze, diversi pesi ogni di paga l'oste* (Of 5 cups of diverse weights to pay the landlord every day), but at the end, Pacioli says that this problem is in "*libro nostro*," i.e., the *Summa*: Cf. Agostini, p. 6.

Pacioli's discussion of perfect numbers has an amusing error (Ff. 44v–47r = Uri 117–122 = Peirani 74–77, *XXVI effecto a trovare un nů pensato quando sia perfecto* (26th effect to find a number thought of if it is perfect)). Pacioli gives the first five perfect numbers as 6, 28, 496, 8128, and 38836. The last is actually $4 \cdot 7 \cdot 19 \cdot 73$ and is so far wrong that I assumed that Peirani had miscopied it, but it is clear in the manuscript. We do have $38{,}836 = 76 \cdot M_{11}$, so it seems Pacioli erroneously thought $M_{11} = 511$ was prime, but the multiplication by 256 was corrupted into multiplication by 76, probably by shifting the partial product by 2 into alignment with the partial product by 5. (See Figure 15.)

The manuscript also includes the first One Pile Game: Ff. 73v – 76v = Uri 175-181 = Peirani 109-112, *XXXIIII effecto afinire qualunch' numero na'ze al compagno anon prendere piu de un termi(n)ato .n.* (34th effect to finish whatever number is before the company, not taking more than a limiting number). The One Pile Game is like Nim, except with just one pile and a limit on the amount one can play. Early versions were usually additive. Here, the players can add a number less than 7 to a pile, and the object

XX VI effecto a trouare un nŭ pen_
sato quando sia perfecto

Vnaltra uolta sopra ognaltra notabilissima.
ci de mostrare' aponto un numero pensato quá_
do lamico dicesse' in questa forma so·ho·un nŭ
dich' quantita siuoglia o·d·o, faue 'cetera el
quale partito in tutti modi integralmente cioe
ch' non uenga rotto ni ancho gli suoi partitore
non sieno con rotti cioe ch' sempre se ragioni co'
sani el numero pensato et li partitori et gli·d'
uenimenti quando cosi sia el amico dica ch'ɔ
tutte ditte parti gionti insiemi fanno apont̃o
el numero quale le ho·pensato dimando ch'ɔ
sia el ditto numerͦ sapi ch' questo non uuol di_
ré altro sinͦ ch' quel tal numero e' perfecto si
comino nel·g̃·el nro phyo de chiara quando cosi
lo disinesa dicendo numero perfecto et quelloch'
aponto atutte le sue parti per li quali se diuide
se aqualia gli quali numeri perfecti de necessitu
sempre siranno terminati in·6·o, uero in·8·alter
natin cioe ch'l primo numero perfecto·et·6·el 2ͦ.
28·el 3ͦ·4 96·el 4ͦ·8128·el·5ͦ· 38 836·et cosi
discorrendo in infinito·cͦ°·de quali numeri per
fecti in tutti modi a pieno nauemo ditto inla

Figure 15. F. 44v = Uri 117 = Peirani 74.

De Viribus Quantitatis by Luca Pacioli

is to achieve 30. Pacioli describes how to win this case and the general game. There are some early sixteenth-century references that may be to Nim-type games, but this is the earliest, so we can view it as the ancestor of all Nim or take-away games! (See Figure 16.)

Pacioli gives several problems of the type "Three odds make an even," which are impossible tricks. These only appear previously in Alcuin. The problems include:

- Ff. 92v–93v = Uri 213-215 = Peirani 132-133, *XLVII. C(apitolo). de un casieri ch' pone in taula al quante poste de d(ucati) aun bel partito* (Chap. 47. of a cashier who placed on the table some piles of ducats as a good trick): Place four piles each of 1, 3, 5, 7, and 9 ducats. Ask the person to take 30 ducats in 5 piles. If he can do it, he wins all 100 ducats. The chapter discusses other versions, including putting 20 pigs in 5 pens with an odd number in each. However, the Italian word for 20, *vinti*, written *uinti*, can be divided into five parts as u–i–n–t–i, and each part is one letter.

- Ff. 93v–94r = Uri 215–216 = Peirani 133-134, *XLVIII. C(apitolo). ch' pur unaltro pone al quante altre poste pare bel partito* (Chap. 48. about another who placed some other even piles, good trick): Place four piles each of 2, 4, 6, 8, and 10 carlini (a small coin of the time) and ask the person to take 31 carlini in 6 piles.

- F. IIIr = Uri 8 = Peirani 6: The index lists the above as Problems 50 and 51 and lists Problem 52, *Del dubio amazar .30. porci in .7. bote disparre* (On the dubious placing of 30 pigs in 7 odd pens).

- Part 3, F. 281v = Uri 591 = Peirani 407, no. 133, *Dimme come farrai a partir vinti in 5 parti despare* (Tell me how to divide "vinti" into five odd parts): It divides as v–i–n–t–i and mentions dividing 20 into 7 pens.

The first optimal solutions for the Jeep or Explorer's Problem, better called Crossing the Desert, also appear. In fact, the only earlier example of this problem is in Alcuin, and Alcuin fails to find the optimum solution. Pacioli does four examples, finding the best solution each time:

- Ff. 94r–95v = Uri 216–219 = Peirani 134–135, *XLIX. (Capitolo) de doi aportare pome ch' piu navanza* (Of two ways to transport as many apples as possible): One has 90 apples to

el suo numero de diuerse cose' acio para piu bello
et a te scusa memoria arteficiale' asseccandote an _
chora el numero dele cose' ch' tu a torno darai se
condo qualch' memorial proportioni commo du _
pla tripla sexquialteria sexquitertia cze acio
tutto te aiutino fra tanti arecordartine' cze
Et da poi Dirai a cada uno ch' prenda altretati
per numero dich' moneta si uoglino ch' uaglia piu
dela prima per membriga et ch' a te di ch' no le
monete prime et seconde ognuno la sua et tu
attenderai ale lor ualute commo disopra e' detto
et simul et semel a tutti aun tratto potrai dire
tu comprafte' tante larance' et tu tanti o, ua et
tu tanti' starne' et tu tanti tordi et tu tanti becha _
fichi ch' sira tenuta una ftupenda cosa maxime
quando con certa graita date simil gentilezze si
ran propofte peroch' tutte gli cose' tanto sono be _
lle' quanto lomo le sa adornare cosi indire commo
in fare ch' tutto la spirientia ci fa chiaro. cze
XXXIIII effecto afinire qualunch' numero na
ze al compagno anon prendere piu de un termiato. h.
Sonno dale predicte forze non da effere' exclusi
alcuni gli giadri guiochi honefte et liciti mathema _
tici quali communa' mente se soliano per li corte

Figure 16. F. 73v = Uri 175 = Peirani 108-109.

De Viribus Quantitatis by Luca Pacioli

transport 30 miles from Borgo [San Sepolcro] to Perosia [Perugia], but one eats one apple per mile and one can carry at most 30 apples. He carries 30 apples 20 miles and leaves 10 there and returns, without eating on the return trip! (So this is the same as Alcuin's version.) Pacioli continues and gives the optimum solution!

- F. 95v = Uri 219 = Peirani 136, *L. C(apitolo). de .3. navi per .30. gabelle 90. mesure* (Of three ships holding 90 measures, passing 30 customs points): Each ship has to pay one measure at each customs point (mathematically the same as the previous).

- F. 96r = Uri 220 = Peirani 136–137, *LI. C(apitolo). de portar .100. perle .10. miglia lontano 10. per volta et ogni miglio lascia 1ᵃ* (To carry 100 pearls 10 miles, 10 at a time, leaving one every mile): This takes them two miles in ten trips, giving 80 there. Then, it takes them to the destination in eight trips, getting 16 to the destination.

- Ff. 96v–97r = Uri 221–222 = Peirani 137, *LII. C(apitolo). el medesimo con piu avanzo per altro modo* (The same with more carried by another method): This continues the previous problem, and it takes them five miles in ten trips, giving 50 there. Then, it takes them to the destination in five trips, getting 25 to the destination.

(The last is optimal for a single stop—if one makes the stop at distance a, then one gets $a(10 - a)$ to the destination. One can make more stops, but this is restricted by the fact that pearls cannot be divided. Assuming that the amount of pearls accumulated at each depot is a multiple of ten, one can get 28 to the destination by using depots at 2 and 7 or at 5 and 7. One can get 27 to the destination with depots at 4 and 9 or at 5 and 9. These are all the ways one can put in two depots with integral multiples of 10 at each depot, and none of these can be extended to three such depots. If the material being transported was a continuous material like grain, then I think the optimal method is to first move 1 mile to get 90 there, then move another 10/9 to get 80 there, then another 10/8 to get 70 there, and so on, continuing until we get 40 at 8.4563..., and then make four trips to the destination. This gets 33.8254 to the destination. Is this the best method?)

Pacioli presents the first impossible Jug Problems: Ff. 98v–99r = Uri 225–226 = Peirani 139–140, *LV. (Capitolo) de doi altri sotili divisioni. de botti co'me se dira* (Of two other subtle divisions of bottles as described). Given a bottle of size A full of wine, divide it in half using two bottles of sizes B and C. After several genuine examples, he gives $\{A, B, C\} = \{10, 6, 4\}$ and $\{12, 8, 4\}$. Pacioli suggests giving these to idiots. This kind of impossible problem is actually rare; I've only noted two other examples.

The first Josephus Problem counted out to the last two is presented in the manuscript. The Josephus, or "counting-out," problem appears in European manuscripts back to the 9th century and also appears in Japan, possibly as early as the 12th century and clearly from ca. 1331. The classical version has 30 passengers on a ship, of two types that we will label "good guys" and "bad guys"—15 of each. A fierce storm arises, and the captain announces that half of the passengers must go overboard to save the ship. Someone suggests that they all stand in a circle and count out every ninth person, who then has to go overboard, willingly or not. After each departure, the count continues, going around the reduced circle. Surprisingly (or not), it happens that all the bad guys go overboard.

The early Japanese versions are the first known examples of counting to the last man. In 1539, Cardan introduced this idea into Europe and suggested this was how Josephus had escaped death. Josephus was a Jewish captain in the rebellion of the Jews against the Romans from 66 AD. He and forty of his fellow citizens were hidden under the city of Jotapata as it was overrun by Vespasian. He urged the men to surrender, but they preferred to die and chose lots, each man striking off the head of the previously chosen man. The standard version of Josephus's text says he survived "by chance or God's providence," but a Slavonic version says he "counted the numbers with cunning and thereby misled them all." Josephus went on to become a historian of the Jews and the Jewish War, but he gives no further details.

Pacioli gives six versions of the problem as Probs. 56–60 (with an unnumbered problem after 56), ff. 99r–103v = Uri 226–235 = Peirani 140–146. In three problems—56, unnumbered, and 57—he leaves two survivors, which is the first time that this occurs. Unusually, there is a marginal diagram by the first problem, showing the process. Ff. 99r–102r = Uri 226–232 = Peirani 140–143, *LVI. (Capitolo) de giudei Chri'ani in diversi modi et regole. a farne quanti se vole etc* (Of Jews and Christians in diverse methods and rules, to make as many as one wants, etc.): two good guys and 30 bad guys counted by 9s. The marginal diagram is on f. 100r, but

el.g. tocco atutti gli 30. giudei et mai 30. uolte re-
uoltandose' sopra gli doi Chriani mai tocco ali
Chriani in modo ch' solo lor doi restaro franchi
inla naue et tutti gli 30. giudei se anegarono di
mandase donde gli Chriani comenzaron acontar'
Dico ch' loro commenzaron a contare. 5. disfatti
da loro. et uenero uerso loro medesimi immo chel
.g. tocco al 2° giudeo a prese gli chrisтiani commo
uedi chi nel cerchio alato nel quale tutti li. O. ne-
ri sonno giudei et li doi rossi sonno chrisтiani et
commenzase dal O pontato. dicendo. 1. 2. 3. ez
et uenendo uerso gli O rossi el. g. la prima uolta
tocco al giudeo 2°. doppo gli chriani cioe al O po'
tato dentro et de fuore' con doi ponti rossi et poi
continuando dal giudeo sequente dal O. trauer
sato. pur dicendo. 1. 2. 3. co el 2. g. tocco al ⊕
incrociato et così continuando la medesima mano
sempre il. g. toccara ali giudei commo e ditto
et mai ali chriani commo per te pornai urmare'
conlo numero deli schachi in un taulieri secon
do el qual numero questo sia formato. cioe fra tutti
32. commo sonno gli schachi et la figura teguida
aponto de mano in mano et se tu domandasse
commo quel chriano si subito al periculo inmi

Figure 17. F. 100r = Uri 228 = Peirani 141 without the diagram.

Figure 18. Enlargement of the diagram in Figure 17.

it is not in the transcription, and Peirani says another diagram is lacking. (See Figures 17 and 18.)

Pacioli suggests counting the passengers on shore and doing the counting out with coins or pebbles, in case one will need to know the arrangement in a hurry. He also says one might count by 8s, 7s, 6s, 13s, etc., with any number of Christians and Jews.

In examining this, I observed the unexpected feature that the two survivors, marked by circles at the top, were adjacent in the original circle. This seemed most unlikely to me, but one soon sees that the same behavior holds for counting out 31, 30, 29, . . . , 3 by 9s. I found this sufficiently intriguing that I have written a paper on how to determine the largest N such that counting out by ks leaves two adjacent survivors.[3]

The other problems, which leave two survivors, have counting 2 and 18 by 7s and counting 2 and 30 by 7s. In the first case, the survivors are adjacent in the original circle, but not in the second case. Neither has a diagram. The problems are:

- Ff. 102r–102v = Uri 232–233 = Peirani 144, [Unnumbered] *de .18. Giudei et .2. Chri'ani.*

- F. 102v = Uri 233 = Peirani 144, *LVII. C(apitolo). de .30. Giudei*

[3]David Singmaster, "Adjacent Survivors in the Josephus Problem."

et .2. contando per .7. ch' toca va in aqua (Of 30 Jews and 2 counting by 7 with the touched going in the water).

The first River Crossing problems with four or more jealous husbands or with larger boats— Ff. 103v–105v = Uri 235–239 = Peirani 146–148, *LXI. C(apitolo). de .3. mariti et .3. mogli gelosi* (About three jealous husbands and three wives)—involves three couples and says that a problem involving four or five couples requires a three-person boat.

Pacioli wrote an early version of the Octagram Puzzle. This has an octagram (or an octagon), and one has to place seven counters on it, by placing each counter on a point and moving it ahead one place (or three places). An earlier version was the problem of shifting seven knights located on the edge of a 3 × 3 board, which is known from ca. 1275. Here are two appearances of the Octagram Puzzle:

- Ff. 112r–113v = Uri 252–256 = Peirani 158–160, *C(apitolo). LXVIII. D(e). cita ch' a .8. porti ch' cosa convi(e)ne arepararli* (Chap. 68. Of a city with eight gates which admits of rearrangement): This is an Octagram Puzzle with a complex story about a city with eight gates and seven disputing factions to be placed at the gates.

- F. IVv = Uri 11 = Peirani 8: The index gives the above as Problem 83. Problem 82, *De .8. donne ch' sonno aun ballo et de .7. giovini quali con loro sa con pagnano* (Of eight ladies who are at a ball and of seven youths who accompany them), seems likely to describe a similar problem.

The first western Binary Divination—Ff. 114r–116r = Uri 256–260 = Peirani 161–162, *C(apitolo). LXIX. a trovare una moneta fra 16 pensata* (To find a coin thought of among 16)—divides 16 coins in half four times, corresponding to the value of the binary digits. Pacioli doesn't describe the second stage clearly, but Agostini makes it clear. This idea is supposed to have been common in Japan from the 14th century or earlier, but I haven't seen examples. Pacioli gives many other simple divinations, some based on the Chinese Remainder Theorem and the classic problem of divining a permutation of three items.

The first Rearrangement on a Cross, a variation of the Blind Abbess and her Nuns, involves a person who has a cross and counts the jewels on it from the base to each other end. A clever jeweler or pawnbroker removes some jewels: Ff. 116r–117v = Uri

260–263 = Peirani 162–164, *Caṗ. LXX. D(e). un prete ch' in pegno la borscia del corporale con la croci de p(er)le al Giudeo* (Of a priest who pledges to a Jew the burse of the corporale with a cross of pearls). For fifteen jewels with three on each arm, one counts to nine from the base to each arm end. This is reduced to thirteen. The problem asks how one can add one pearl and produce a count of ten. The answer is to put it at the base. (See Figure 19.)

Pacioli associates magic squares with planets and gives Dürer's magic square of 1514, but both of these had been done before: Ff. 118r–118v, 121r–122v (some folios are wrongly inserted in the middle) = Uri 264–265, 270–273 = Peirani 165–167, *C.A. [i.e., Capitolo] LXXII. D(e). Numeri in quadrato disposti secondo astronomi ch' p(er) ogni verso fa'no tanto cioe per lati et per Diametro figure de pianeti et amolti giuochi acomodabili et pero gli metto* (Of numbers arranged in a square by astronomers, which total the same in all ways, along sides and along diagonals, as symbols of the planets and suitable for many puzzles and how to put them). The problem gives magic squares of orders 3 through 9 associated with planets in the system usually attributed to Agrippa (1533), but this dates back to at least the early fouteenth century. Ff. 121v and 122r have spaces for diagrams, but they are lacking. Paciloi gives the first two lines of the order-4 square as 16, 2, 3, 13 and 5, 10, 11, 8, which is the same square as given by Dürer.

He gives an example of "Selling different amounts 'at the same prices' yielding the same": Ff. 119r–119v = Uri 266–267 = Peirani 154–155, *LXV. C(apitolo). D dun mercante ch' a .3. factori et atutti ma'da auno mercato con p(er)le* (Of a merchant who has three agents and sends them to a market with pearls).

In addition, there are four more examples listed in the index, Ff. IIIv–IVr = Uri 9–10 = Peirani 7, as problems 70–73:

- Problem 70: *De unaltro mercante ch' pur a .3. factori et man-dali a una fiera con varia quantita de perle' et vendano a mede-simo pregio et portano acasa tanti denari al patrone uno quanto laltro* (Of another merchant who sends three agents to a fair with varying numbers of pearls and they sell them at the same price and they each carry as many pence as the others to the master at home).

- Problem 71: *De unaltro vario dali precedenti ch' pur a .3. fac-tori con vari quantita de perle' pregi pari et medesimamente portano al patrone d(enari) pari* (Of another variant of the pre-ceding with three agents having various quantities of pearls

la 3.ª cioe. c. peroch' questi .z. ultimi fili l'uno
sira. g. et. c. a. q. o. m. k. l'altro sia. h. f. d. b.
. r. p. n. l. et cosi obseruami sempre. et date
piu altre. ne proportionarai c͛.

     C Ap̃. Lxx. D. un prete ch' in pegno la
     borscia del corporale con la croci de ple'
         al Giudeo

Q Vuanto facia al huomo acolto scatrito la forza et uirtu deli numeri oltra legia asegnate
occurrentia' el fa manifesto el caso ch' gia a-
un certo poucano interuene el quale accerti
suoi bisogni no' hauendo altri denari ardo
al giudeo a' impegnare una bella ueste' de
corporale de grmn ualuta per una certa q̃
de dp̃. sopra la quale eranno al quante gro-
sse perle destima situate in croci la qual croci
era in questo modo facta commo uedi qui in
margine et cioe ch' contandole al piedi cioe da-
piedi al capo eranno noue perle. et gli piedi
con l'un de braci anchora. g̃. et cosi gli piedi
con l'altro bracio simil mentz g̃. et lasciandola
al giudeo acio non si scambiasse' ne prese el
ditto contrase deman del giudeo cioe ch' da lui
receue una uesta tale c͛ con una coci tale

at equal prices and likewise taking as many pence to the master).

- Problem 72: *De unaltro mercante ch' ha 4. factori ali quali da quantita varie di perli ch' amedisimi pregi le vendino et denari equalmente portino* (Of another merchant who has four agents to whom he gives various numbers of pearls, which they sell at the same prices and receive equal money).

- Problem 73: *De un altro ch' pur a .4. factori con quanti(ta) varie di perle apari pregi et pari danari reportano a casa vario dali precedenti* (Of another who sends four agents with varying numbers of pearls, and they report back to the house the same prices and the same money, variation of the preceding).

He gives an example of "Combining amounts and prices incoherently," sometimes called the Applesellers' Problem or the Marketwomen's Problem: Ff. 119v–120r = Uri 267–268 = Peirani 155–156, *LXVI. C(apitolo). D. de uno ch' compra 60. perle et revendele aponto per quelli ch' gli stanno et guadag°* (Of one who buys 60 pearls and resells for exactly what they cost and gains). The solution is to buy 60 at 5 for 2, sell 30 at 2 for 1, and sell 30 at 3 for 1.

The index, F. IVr = Uri 10 = Peirani 7, lists the above as Problem 74 and continues with Problem 75: *De unaltro mercante ch' pur compro perle' .60. a certo pregio per certa quantita de ducati et sile ceve'de pur al medesimo pregio ch' lui le comparo et guadagno un ducato ma con altra industria dal precedente* (Of another merchant who buys 60 pearls at a certain price for a certain quantity of ducats and resells them at the same price at which he bought them and gains a ducat but with different effort than the preceding).

The problem of gathering apples from a garden appears in Ff. 120r–120v, 111r–111v (some pages are misbound here) = Uri 268–269 & 250–251 = Peirani 156–158, *C(apitolo). LXVII. un signore ch' manda un servo a coglier pome o ver rose in un giardino* (A master who sends a servant to gather apples or roses in a garden), which involves losing half and one more three times to leave one. The author discusses the problem in general and also discusses losing half and one more five times to leave one and losing half and one more three times to leave three.

Pacioli gives an example of Collecting Stones, which is a simple summation of an arithmetic progression: Ff. 122v - 124r = Uri 273-276 = Peirani 167-169, *C(apitolo) LXXIII. D(e). levare .100. saxa a filo* (To pick up 100 stones in a line). One wagers on the number of steps to pick up 100 stones (or apples or nuts), one pace apart. The problem gives the numbers for 50 and 1000 stones.

Pacioli gives rules for constructing (approximate) n-gons, for n = 9, 11, 13, and 17, which were studied by Mackinnon.[4] Let L_n denote the side of a regular n-gon inscribed in a unit circle. The rules given are:

- Ff. 147r–147v = Uri 322–323 = Peirani 198–199, *XXIII afare la 7ª fiª dicta nonangolo. cioe de .9. lati difficile* (23 to make the 7th figure called nonagon, that is of 9 sides, difficult): This problem asserts that $L_9 = (L_3 + L_6)/4$. Mackinnon computes this as giving 0.6830 instead of the correct 0.6840.

- Ff. 148r–148v = Uri 324–325 = Peirani 200, *XXV. Documento della 9 fiª rectⁱ detta undecagono* (25 on the 9th rectilinear figure called undecagon): This asserts that $L_{11} = \phi\,(L_3 + L_6)/3$, where ϕ is the golden mean, $(1 + \sqrt{5})/2$. Mackinnon computes this as giving 0.5628 instead of the correct 0.5635.

- F. 148v = Uri 325 = Peirani 200, *XXVI. Do. de' .13.* (26 on the 13th). This asserts that $L_{13} = (1 - \phi) \cdot 5/4$. Mackinnon computes this as giving 0.4775 instead of the correct 0.4786.

- Ff. 149r–149v = Uri 326–327 = Peirani 201–202, *XXVIII. Documento del .17. angolo cioe fiª de .17. lati* (28 on the 17-angle, that is the figure of 17 sides). Peirani says some words are missing in the second sentence of the problem, and Agostini says the text is too corrupt to be reconstructed.

In Part 2, Pacioli presents the first Staircase Cut: Ff. 189v–191r = Uri 407–410 = Peirani 250–252, *LXXIX. Do(cumento). un tetragono saper lo longare con restregnerlo elargarlo con scortarlo* (to know how a tetragon can be lengthened with contraction, enlarged with shortening).

I was looking at this problem a few days before the G4G5 talk, since it has an added diagram, seems to use a trick cut, and might be an ancestor of the vanishing area puzzles. Pacioli's description is a little cryptic and is thoroughly confused by an erroneous diagram added at the bottom of f. 190v, redrawn on Peirani 458—this must have been added by a reader who didn't understand the phrasing.

---

[4]Nick Mackinnon, "The Portrait of Fra Luca Pacioli," *The Mathematical Gazette* 77: 479 (July 1993), plates 1–4 and pp. 129–219.

Once one realizes what is going on, the text is reasonably clear. He is converting a 4 × 24 rectangle to a 3 × 32 using one cut into two pieces. So, this is the common problem of converting from $4A \times 3B$ to $3A \times 4B$, with $A = 1$ and $B = 8$, which is done by a "staircase" cut, giving two pieces that can be assembled into a second rectangle. Below the diagram on f. 190v is an inserted note, which Peirani (252) simply mentions as difficult to read, but some bits of it are legible. The drawing and the note made me think he made a cut and then moved one piece so the cut would continue through it to make three pieces with one trick cut. Pacioli clearly notes that the area is conserved. (See Figures 20 and 21.)

The first Place Four Points Equidistantly problem is described, though a bit vaguely: Ff. 191r–192r = Uri 410–412 = Peirani 252–253, *LXXX. Do(cumento). commo non e possibile piu ch' tre ponti o ver tondi spere tocarse in un piano tutti* (how it is not possible for more than three points or discs or spheres to all touch in a plane). This problem says that you can only get three discs touching in the plane, but you can get a fourth so that they are all touching by making a pyramid.

Pacioli gives the earliest known versions of six "topological puzzles." Unfortunately, only one of these has a picture, though they generally refer to one! I had recognized some of these, but Dario Uri has greatly extended the number of these. One is the first Victoria Puzzle, called the Alliance or Victoria Puzzle in the late nineteenth century: Ff. 206r–206v = Uri 440–441 = Peirani 282–283, *(C)apitulo. C. Do(cumento) cavare una stecca. de un filo per .3. fori* (To remove a stick from a cord through 3 holes). On f. 206r is a marginal drawing clearly showing the string through three holes in one stick, but this is not reproduced in the transcription. (See Figures 22 and 23.)

Uri has found that several further problems are describing similar puzzles. For example,

- Ff. 207v–208v = Uri 443–445 = Peirani 284–286, *Capi°. CII. Do(cumento) unaltro speculativo cavar doi botoni di una stenga fessa nel mezzo et sce'pia in testa* (Another speculation— remove two buttons from a string divided in the middle and halved at the ends). Dario Uri says this is describing a version of the Alliance or Victoria puzzle with four holes in each button.

- Ff. 209r–210r = Uri 446–448 = Peirani 286–288, *Cap°. CIII. Do(cumento) legare con la sopra detta strenga fessa. doi sola. de carpe' ambe doi. a uno modo. bella cosa* (To tie two shoe

superficie delati equi dustanti slongarla con
suo restregnerla. cioe farla piu stretta et cosi
largarla con farla piu corta la qual cosa
in uno sol taglio si fa cioe con resoluerla
solo in doi pezzi commo gia stando io infe
rrara. nel 1466. per la festa de sangiorgio
la exca del Duca borso uolendo far correr
el magno palio de brocato doro quale secodo
lusanza bisognaua fosse longho una certa
quantita de braccia elargo un tanto troue
un mercante ch' nauiua un pezzo longho
z 4. elargo 4. elui el uoliua longo 3 z
elargo 3. et pareua strano auederlo cosi fi
naliter fochi il taglio in doi pezzi non guas
tando orli emendato con lago gentilmente fo
al bisogno cioe largo 3. elongo 3 z. se doma
da commo fo dicte taglio. Dirai ch' fo como
uedi qui la prima dispositione cioe ch' lo taglio
dustante dalun canto. uerso. 1. commo in dicta
figura a b c d. lontano dalangolo. a. i.e et uene
giu pel longo. 8 X. cioe fin f. equi uolto gli
forbici. a trauerso tagliando fin g. cioe X. i°.
commo a e. et poi le uolto per lo longho tagli
ando fin h. cioe altri. X. 8. quanto e.f. epoi

q. 10.

Figure 20. F. 190r = Uri 408 = Peirani 251.

A Lifetime of Puzzles

epoi le uolto atrauerso tagliando fin lz. cioe
unaltro bracio Et poi le uolto per lo longo Are—
gnendole fin.l. cioe altri 8. ch' son doi pezzi
aponto et poi le sopra mise' insu quelle tach
commo uedi in la z.ª dispositione ch' la tacha
del f. ando in su lo.a. et quella delo h. insu
f. et quella de.l. in sulo h. elangolo.c. se alzo
8. et cosi·e· sopra a. in modo ch' resto 3. lar—
ga et douento longa 3z. cioe crebe 8. et callo
l'. o, uoi dire crebe el ⅓ ecallo el 4º. ch' tanto
uale et non se uerde niente de superficie pero
ch' prima era 9 6. cioe 4. uerra z4. et anchora
mo pur 9 6. cioe·3. uerra 3z. et cosi se fosse
longa 1z. elarga 8. afarla longa 16. e larga
·6. taghara 1. in sul·z. della larghezza per
lo longo andarai in sul ⅓. della longhezza cioe
4. e poi atrauerso pur z. et poi alengiu per
lo longo 4. epoi atrauerso z. et poi pel lon —
go fin al fondo et sopra porrai le tacch cómo
uedi in la 3.ª. et 4.ª. dispositione et cosi se fo —
sse longo 9 6. elargo·1. et tu uolesse largo z.
et longo 4 8. taglia a trauerso in la ½ et a—
costala asiemi. harai el bisogno per ch' calla
la ½. et cresci altre tanto ∴ E cosi se fosse

206

APITVLO. C. Do cauare una stecca. de un filo
—. per 3. fori:—
Sonno alcune operationi facte per dar dilecto alla bri
gata quali sonno de grande speculatione. et fan si fiali
genuani per acomodare loro ingegni a maggiore facti
dele quali luna fin questa. cioe una stecca. ha. 3. fori
qual sia la qui figurata. a b. et li suoi fori luno. c. lal
tro. d. laltro. e. et per questi fori io metto. un filo do
pio. in questo modo. ch'la capia metto in un de quelli
de le teste et sia in quello. e. de su in giu et quella me
desima capia ne metto. desotto in su nel foro d. in lo
mezzo poi in questa capia qual sia. f. io metto lal—
tro capo del ditto filo dopio et questo medesimo capo
ni metto desu in giu in lo terzo foro. c. e tengo lo sal
do in mano lasciando la stecca. cosi ligata etessuta
in habandono. in liberta del compagno. acio la possi
a su libito manegiare non mouendo el mio capo de
mano senzale compere de filo o stecca. dimandase co
mo se cauara standoipossibile. Dirai il caso esser possi
bile et fasse in questa forma. u z. prendi ditta capia
f. e quella farai passare. per lo foro c. de su in giu cioe
per lo foro doue me tesa. el capo ch' tieni in mano a lo
tando el filo in modo ch'tu lo possi alibito manegiare
estendere et quando la rai facta passare per ditto foro

Figure 22. F. 206r = Uri 440 = Peirani 282-283, without the diagram.

A Lifetime of Puzzles

Figure 23. Detail from Figure 22.

soles together into one with the above mentioned doubled string—a beautiful thing). Pacioli says this is "*quasi simile alla precedente.*" Dario Uri illustrates this with the Alliance or Victoria puzzle from Alberti; see Figure 24.

- Ff. 213r–215r = Uri 454–458 = Peirani 292–295, *Capitulo. (C)VIII. Do(cumento). Cavare' uno anello grande fore' de doi legati a una bacchetta per testa* (Remove a large ring from two tied to a stick by the ends). Dario Uri says there is a version of this idea and illustrates it with an unidentified picture.

Part 2 also includes the first Solomon's Seal (see Figure 25): Ff. 206v–207r = Uri 441–442 = Peirani 283–284, *Capitulo. CI. Do(cumento) un altro filo pur in .3. fori in la stecca con unambra. per sacca far le andare' tutte in una* (Another string also through three holes in the stick with one bead per loop, make them go onto one [loop]). The problem titles vary between the actual problem and the table of contents, and the latter shows that "*unambra*" should be "*una ambra*"—Peirani has given it as *un'ambra. Sacca* means "pocket" or "bay" or "inlet," and it seems clear he means a loop that has that sort of shape. *Ambra* is *amber*, but it seems to mean an amber bead here. In recent years, this has been called an African puzzle, but the earliest recorded appearances in Africa are from ca. 1940.

The Cherries Puzzle has two versions, and Pacioli gives both:

- Ff. 210r–210v = Uri 448–449 = Peirani 288–289, *Capit°. CIIII. Do(cumento). cavare' et mettere' .2. cirege' in una carta tramez-zatta* (To remove and replace two cherries in a cut card): Pacioli's description clearly shows there is one hole, but Dario

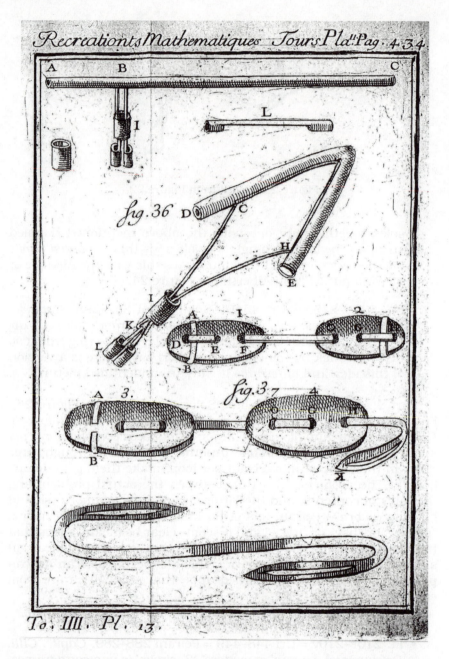

Figure 24. The lower puzzle is the Ozanam version of ca. 1723, later copied by Alberti.

gung mögen referirt werden/mir keins wunderſamer vorkommen/als diß/
obs zwar bey den Wiſſenden ein ſchlecht anſehen hat/wolte wündſchen/daß
ich die demonſtration alſo dazu ſetzen könnte / daß ſie von mäniglich möch=
te verſtanden werden/weil ſie aber allzulang vnd mühſam/will ich den gün=
ſtigen Leſer damit nit moleſtiren oder beſchweren/ſondern einig vnd allein/
wie man hierinn practicire,jhme an die Hand geben. Jch halte dafür/daß
niemand von ſich ſelbſt/beede Ring dem begeren nach/zuſamm bringen wer=
de: Das Holtz aber dazu rd alſo gemacht: Nimb ein Holtz vngefehr eines
Meſſerrucks dick/vnd einer ſpannen lang / ſpitze es zu wie bey der Figur im

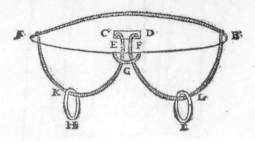

Figure 25. Solomon's Seal from Schwenter, 1636.

Uri illustrates this with the picture from Alberti that has two
holes. (See Figure 26.)

- Ff. 215v–216v = Uri 459–461 = Peirani 296–297, *Capitulo.
  CX. Do(cumento). uno bottone' de un balestro. o vero doi cirege'
  de un botone' et valestro* (A button from a [cross] bow or two
  cherries from a button and bow). Dario Uri translates "*balestro*" as "flexible stick" and illustrates this with Alberti's Figure 37.

See Figure 24 above for the second form of the cherries puzzle as
the upper part of this figure from Ozanam, ca. 1723, later copied
by Alberti.

Pacioli describes (without a picture) the first Chinese Rings puzzle in Europe (or in the world?): Ff. 211v–212v = Uri 451–453 =
Peirani 290–292, *Capitulo CVII. Do(cumento), cavare et mettere una
strenghetta salda in al quanti anelli saldi. dificil caso* (Remove and
replace a joined string with a number of joined rings—a difficult
thing). Dario Uri found that this describes the Chinese Rings. It
has seven rings. Previously, the earliest known version was given

XIX.

Een ander.

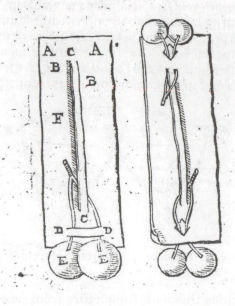

Figure 26. Cherries puzzle from Witgeest, 1686.

by Cardan in his *De subtiliate* of 1550, where his only illustration is of one ring! On his website, Uri gives several of the legends about its invention and says that Cardan called it *Meleda*, but that word is not in Cardan's text. He lists 27 patents on the idea in five countries. It is supposed to have originated in China, but definite evidence is lacking. (See examples in Figure 27.)

Also included is the first puzzle involving three knives making a support: Ff. 228r–228v = Uri 484–485 = Peirani 315, *Capitolo. CXXIX. Do(cumento). atozzare .iij tagli de coltelli insiemi* (Join together three blades of knives). Pacioli says that this was shown to him on April 1, 1509 (Peirani has misread "1509" as "isog") by "*due dorotea veneti et u perulo 1509 ad primo aprile ebreo.*" Peirani transcribes "*u*" as "*un*," but Dario Uri thinks it is the initial of Perulo's given name. I wonder if "*dorotea*" might refer to some occupation, e.g., nuns at St. Dorothy's Convent. In Vienna, the Dorotheum is a huge public auction house where estates are auctioned off. The

A Lifetime of Puzzles

九连环和巧环——由阮根全制作
Ingenious Rings Puzzles by Ruan Genquan

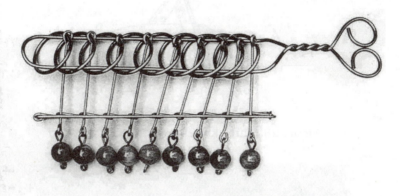

九连环—要将框柄从所有九个连环中套出需 341 步
Nine Interlocking Rings is a classical Chinese puzzle.
Removing the handle from all nine rings requires 341 moves.

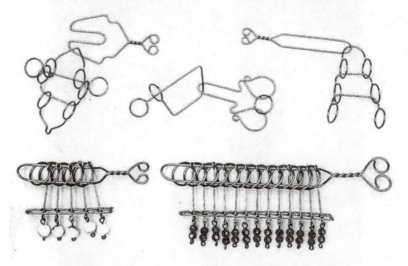

双套连环、剪和刀环、十全环、五连环、十三连环
Five Interlocking Rings,Thirteen Interlocking Rings,and other puzzles

Figure 27. Some modern examples of Interlocking Rings from China.

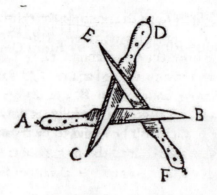

Figure 28. Three knives from Prevost, 1584.

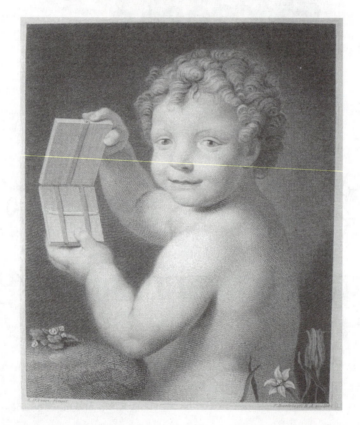

Figure 29. A 1793 engraving by Bartolozzi of a painting by Luini, ca. 1510.

A Lifetime of Puzzles

word "*ebreo*" means "Hebrew," but I cannot see what it refers to. (See Figure 28.)

We also find in Pacioli's work the first mention of Jacob's Ladder (or Flick-Flack). The early form with two boards is also known as the Chinese Wallet. Versions with more boards appear in the late eighteenth century: Ff. 229r–229v = Uri 486–487 = Peirani 316–317, *Capitolo. CXXXII. Do(cumento). del solazo puerile ditto bugie* (On the childish recreation called deception[?]). It uses two tablets and three leather straps. The problem describes how to use it to catch a straw. Peirani's references show that it is called "*calamita di legno*" in Italian ("calamity of wood" or "magnet of wood," depending on whether the Italian is "*calamità*" or "*calamita*"). (See Figure 29.)

Part 3 includes the first discussion of a geometrical optical illusion: Ff. 256v–257r = Uri 541–542 = Peirani 364–365, *Capitolo. LXXIII. Do. in gannare' uno della vista abagliarlo* (to deceive someone's eyes, an illusion). It takes two identical strips of paper and places one perpendicular to the other to make a T shape. Nine out of ten people say one direction is longer than the other. Then he interchanges the sheets, but the same direction is still seen as longer. This is generally called the *vertical-horizontal illusion* and is attributed to Oppel (1855) or Fick (1851). (Thanks to Vanni Bossi for pointing out this item.)

The third part also includes the first Two Fathers and Two Sons Make Only Three People puzzle: F. 287v, no. 191 = Uri 603 = Peirani 416. Pacioli gives several other "strange family" puzzles.

Vanni Bossi and Bill Kalush have looked at the tricks and discovered several earliest examples of magic tricks; see Bossi's article immediately following this one. Those who have looked at this manuscript now feel that it is definitely the earliest recreational mathematics book—except that it was never published.

Borgo San Sepolcro has recently commemorated Pacioli on the occasion of the 500th anniversary of the *Summa*, both with a handsome facsimile of it and a statue of him (Figure 30).

Acknowledgments. My thanks to Dario Uri and Bill Kalush.

Appendix 1: Chronology

ca. 1445 Born in San Sepolcro.

1466–1470 Pacioli spent some time in Venice, teaching in the house of Antonio de Rompiasi in the Giudecca, near Sant' Eufemia,

Figure 30. Statue of Luca Pacioli in Borgo Sansepolcro.

and attended lectures of Domenico Bragadino at the Scuola di Rialto.

1470+ Pacioli stayed with Alberti in Rome for about a year. He later became attached to Cardinal Francesco della Rovere and lived at his palace adjacent to the basilica of St Pietro in Vincoli. The Cardinal became Pope Sixtus IV in 1471, and his nephew Giuliano became a cardinal (later Pope Julius II) and took over the palace and Pacioli.

ca. 1475 Pacioli entered the Franciscan Order in San Sepolcro.

1477–1480 Luca Pacioli was the first mathematics lecturer at the University of Perugia. His biographer says that Pacioli states in his *Summa* that he was here in 1475–1480 and that in 1475, he lectured to a class of 150.

In the Galleria Nazionale dell'Umbria is another of Piero della Francesca's paintings: *The Polyptych of St. Antony* of 1470. Pacioli's biographer claims that the leftmost saint "is the same as" the St. Peter Martyr in the picture in the Brera, Milan, and hence is a picture of Pacioli, but this is dubious.

1481–1486 Pacioli taught at Zara, on the Dalmatian coast and then part of Venice.

1486–1488 Pacioli was lecturing at the University of Perugia.

1489 Pacioli was in Rome, at the palace of Cardinal Giuliano della Rovere.

1490s Pacioli taught in Naples for some time, possibly completing his *Summa* here, though he certainly worked on it while it was being printed in Venice. There is a passage in *De viribus quantitatis* where Pacioli refers to one of his pupils in 1486 in Naples, but this may have been a pupil in 1486 who was later in Naples.

1493 Pacioli gave lectures on mathematics in San Sepolcro. Dürer came to Venice during his Wanderjahre in 1493 and knew Jacopo de Barbari, painter of the famous portrait of Pacioli, now in Naples.

1494 Pacioli came to Venice to publish his *Summa*. As a Franciscan, he stayed at the Ospite del Convento dell'Ordine al Ca'Grande.

1497 Pacioli contracted to have a chapel of St. Bernardino built in the Church of St. Francis in San Sepolcro and is described as Warden of the Monastery of St. Francis. Another 1497 document describes him as a Professor of Holy Writ.

1496–1499 Pacioli is Professor at Milan. He was inspired to start his *Divina Proportione* on February 9, 1498 and completed it on December 14, 1498, though it was not published (in an expanded form) until 1509.

1499–1507 Teaching at the Universities of Florence and Pisa, living in the monastery of Santa Croce. He and da Vinci had left Milan together and came to Florence, originally lodging in the same house. He spent most of 1500–1507 here.

1508–1509 Pacioli returned to Venice to publish his *De Divina Proportione* and gave a lecture on the Fifth Book of Euclid at the Church of S. Bartolomeo on August 11, 1508.

ca. 1510 He retired to San Sepolcro as Commissioner (head or warden) of the Franciscan House, though he went briefly to teach in Perugia in 1510 and in Rome in 1514.

1510 Pacioli taught at Perugia.

1514 Pacioli was in Rome, at the palace of Cardinal Giuliano della Rovere.

1517 Pacioli probably died in San Sepolcro, but there is no gravestone or grave site.

There is a street named after him and there is a plaque with his portrait on the Palazzo delle Laudi (photos of the Palazzo and the plaque are in Giusti, pp. 20–21). A statue of Pacioli was recently erected in the park at Via Matteotti and Via de gli Aggiunta. The base uses designs from his *De Divina Proportione*. The leading hotel/restaurant in town is the Albergo Ristorante Fiorentino at Via Luca Pacioli 60, and the proprietor is interested in Pacioli: in the restaurant is a banner from the 500th anniversary celebrations of the publication of *Summa* in 1494.

In San Sepolcro, Piero's house is at the corner of Via Piero della Francesca and Via de gli Aggiunti. Piero is buried in the Cathedral. There is a statue of him. The Museo Civico has a portrait and a bust of him.

Appendix 2: Works

ca. 1480? Piero della Francesca's manuscript *Trattato d'Abaco*. Italian manuscript in Codex Ashburnhamiano 359* [291*] - 280 in the Biblioteca Mediceo-Laurenziana, Florence.

This work and Piero's *Libellus de Quinque Corporibus Regularibus* of ca. 1487 are the subject of a long-standing plagiarism argument. Giorgio Vasari, in his *Le Vite de' più eccellenti pittori, scultori e architetti* of 1550, states:

> ... Piero della Francesca, who was a master of perspective and mathematics but who first went blind and then died before his books were known to the public. Fra Luca di Borgo, who should have cherished the memory of his master and teacher, Piero, did his best, on the contrary, to obliterate his name, taking to himself all the honour by publishing as his own work that of that good old man. ... Maestro Luca di Borgo caused the works of his master, Piero della Francesca, to be printed as his own after Piero died.

The mathematical works of Piero were unknown until they were rediscovered in 1850/1880 and 1917. Examination shows that Pacioli certainly used 105 problems, many unusual, from Piero in the *Summa*. But he does praise Piero in the *Summa*, as "the monarch of painting of our times." Entire books have been written on the question, so I will not try to say any more.

ca. 1487 Piero della Francesca. *Libellus de Quinque Corporibus Regularibus.*

Piero would have written this in Italian, and it is believed to have been translated into Latin by Matteo da Borgo [Davis, p. 54], who improved the style. It is the first post-classical discussion of the Archimedean polyhedra, but it was not published until an Italian translation (probably by Pacioli) was printed in Pacioli and da Vinci, q.v., in 1509, as *Libellus in tres partiales tractatus divisus quae corpori regularium e depēdentiū actine perscrutatiōis . . .*, ff. 1–27. A Latin version was discovered by J. Dennistoun, ca. 1850 and rediscovered by Max Jordan, 1880, in the Urbino manuscripts in the Vatican—manuscript Vat. Urb. lat. 632.

Davis identifies 139 problems in this, of which 85 (61%) are taken from the *Trattato*. The Latin text differs a bit from the Italian.

Piero describes a sphere divided into 6 zones and 12 sectors. He gives the truncated tetrahedron, truncated cube, truncated octahedron, truncated dodecahedron and truncated icosahedron—see below for the cuboctahedron—and there is an excellent picture of the truncated tetrahedron on f. 22v of the printed version. The Latin manuscript gives different diagrams than in the 1509 printed version, including clear pictures of the truncated icosahedron and the truncated dodecahedron. An Internet biographical piece, apparently by, or taken from, J. V. Field,[5] shows that the counting is confused by the presence of the cuboctahedron in the *Trattato* but not in the *Libellus*. So della Francesca rediscovered six Archimedean polyhedra, but only five appear in the *Libellus*.

1494 *Summa de Arithmetica, Geometria, Proportioni et Proportionalità*, Venice, 1494. This is a massive book, 616 large pages, too large for my scanner! A facsimile was produced in 1994.

Part II, ff. 68v 73v, prob. 1–56, are essentially identical to Piero della Francesca's *Trattato*, ff. 105r–120r.

---

[5]http://www-history.mcs.st-andrews.ac.uk/history/Mathematicians. Francesca.html

1498 Pacioli and da Vinci, *De Divina Proportione*. The manuscript begins with "*Tavola dela presente opera e utilissimo compendio detto dela divina proportione dele mathematici discipline e lecto.*" Three copies of this manuscript were made. One is in the Civic Library of Geneva, one is the Biblioteca Ambrosiana in Milan, and the third is lost. Modern facsimiles exist. It contains illustrations of six Archimedean polyhedra and the first Stella Octangula.

ca. 1500 *De Viribus Quantitatis*. Italian manuscript in Codex 250, Biblioteca Universitaria di Bologna. Pacioli petitioned for a privilege to print this in 1508 and a problem has a date of 1509, but he seems to have been working on the manuscript since 1496. The title is a bit cryptic, but I think the best English version is *On the Powers of Numbers*.

1509 Pacioli and da Vinci: *[De] Divina proportione Opera a tutti glingegni perspicaci e curiosi necessaria Ove ciascun studioso di Philosophia: Prospectiva Pictura Sculptura: Architectura: Musica: e altre Mathematice: suavissima: sottile: e admirabile doctrina consequira: e delectarassi: cōvarie questione de secretissima scientia.* Illustrations by Leonardo da Vinci. Includes Piero della Francesca's *Libellus* and other extra material. Paganino de Paganini, Brescia, 1509. Modern facsimiles exist.

The printed version was assembled from three codices dating from 1497–1498 and contains the 1498 manuscript, with several additional items.

Pacioli and da Vinci give six Archimedean solids. They assert that the rhombicuboctahedron arises by truncating a cuboctahedron, but this is not exactly correct.

Part of the printed version is *Libellus in tres partiales tractatus divisus quae corpori regularium e depēdentiū actine perscrutatiōis* ..., which is an Italian translation (probably by Luca Pacioli) of Piero della Francesca's *Libellus de quinque corporibus regularibus*. Some architectural material, and the handsome and often reproduced geometric designs for letters of the alphabet, are also appended.

Davis says the drawings were made from models prepared by da Vinci, but Pacioli made, or had made, at least three sets of 60 models.

Magic and Card Tricks in Luca Paciolo's *De Viribus Quantitatis*

Vanni Bossi

The *De Viribus Quantitatis* manuscript is a collection of "*ludi mathematici*" (mathematical games or recreations) where the author wishes to teach mathematics and avoid the weariness of repeated exercises that normally ask for the power of intellect and patience.

For the same reason, other authors before him did the same (like Leonardo Pisano, better known as Fibonacci, or both Francesco and Pier Maria Calandri). But, in other "*trattati d'abbaco*" (abacus treatises, which were actually arithmetic textbooks; both *abbaco* and *abaco* occur), recreational problems are placed here and there in the text, just to give a "pause" to the mind, while Paciolo's manuscript can instead be considered a real treatise on the subject.

This article describes Bossi's part of a joint presentation between him and David Singmaster at the sixth Gathering for Gardner, 2004. Singmaster's part immediately precedes this article.

The relationship with magic is clearly understandable in some of Paciolo's statements. For instance, he gives great importance to the secret of the method used to accomplish the effect to astonish the viewers, which is a basic and fundamental principle in magic. His great care in hiding the secret is, he says, because the secret is the "*conditio sine qua non*" to be able to amaze your friends, especially "*i rozzi*" (the rude fellows) and "*maxime donne*" (especially women), who don't know mathematical principles because neither had access to the "*scuole d'abaco*" (abacus schools or schools of arithmetic). So, if some rules are easy to learn and master, some others are intentionally complicated, augmenting the number of necessary operations to avoid detection. This allows you both to give a different presentation of the same effect, or to disguise the method using a different one, which is a second fundamental principle greatly used in magic. Further, in most cases, the spectator is invited to simply think of a number or the value of a coin or card, which adds to the mystery.

Luca Paciolo's (or Pacioli's, or Fra Luca da Borgo Sansepolcro's, or "fra Luca's") manuscript has remained unpublished for about 500 years. We could consider the Peirani Marinoni transcription as the first printed edition of this work.[1] Notwithstanding this, we have much proof that it has been a source for later works. One popular work is Bachet's *Problemes plaisantes et delectables* to which, in the past, various scholars attributed the priority for being the first work on recreational arithmetic. Bachet's respected work is the first printed book, but the honor of producing the first collection of entertaining mathematical problems belongs to Paciolo.

Fra Luca doesn't claim originality; some of the games come from older works. Some have been invented by his students, and he gives proper credit. (Paciolo explains that he encouraged his students to do this.) For instance, in Chapter XLVII, he names his disciple Carlo de Sansone from Perugia; in Chapter XLVIII, he names Catano de Aniballe Catani from Borgo, who performed the game in Naples in 1486. This date is interesting because it suggests that most of the problems in the manuscript could have been invented in the last quarter of the fifteenth century.

According to Gilberto Govi, one of the great nineteenth-century scholars of Leonardo, most of the tricks should be credited to Leonardo. This is absolutely possible, as Fra Luca taught Leonardo arithmetic and Euclid's geometry, although unfortunately the man-

---

[1] Luca Pacioli, *De viribus quantitatis*, Milano, Ente Raccolta Vinciana, 1998.

uscript doesn't give credit to Leonardo for any of them. It is known that two "*giochi di partito*" (party games) are described in notebooks of Leonardo, so this could be a proof of the connection. Govi also wrote that Leonardo performed a trick where some light objects made of wax (I believe, probably tissue waxed to make it stiff and moldable) were made to fly in the way that oriental performers make paper butterflies fly! Again this is possible. Soon after the death of Professor Augusto Marinoni, Giunti Editore in Florence published a monumental work that is his complete transcription of Leonardo's *Codice atlantico*. I met Professor Marinoni many times (he was living in Legnano, very close to where I live), and he confirmed to me that Leonardo had an interest in conjuring and performed some tricks. Unfortunately, the personal notes and files of Professor Marinoni are unpublished and unavailable at present.

Back to Paciolo, we can also suppose that he shared some of his secrets with street conjurers or court performers; this would explain his knowledge of most of the non-arithmetical tricks and puzzles described in the second and third parts of the manuscript. This theory could also explain the finding of some principles explained in the manuscript, which we know was never printed and was hardly accessible to common people, in many pamphlets of "secrets," usually sold by itinerant performers. These "secrets" were probably transmitted orally and occasionally printed.

Many of these booklets have been discovered recently. As far as is presently known, the most extensive work is Horatio Galasso's *Giochi di carte bellissimi, di regola e di memoria ...* published in Venice, 1593, in which the author describes 25 card tricks (including the first printed system for a stacked deck), many of them based on arithmetical principles, some of which can be found in Paciolo's work. This is followed by 25 "secrets" of various types, some of which can also be found in *De Viribus Quantitatis*. In 2001, I made a reprint of this booklet, with an introduction. It has been translated into English, and I hope it will soon be available to English-speaking people.

Coins, dice, and cards are the objects most used in the explanation of tricks based upon arithmetical principles, the reason being that all these objects can represent a number or quantity: an amount of equal coins with each one a unity, or coins with different values; dice with six faces of different values and the possibility to use more than one of them; and playing cards with values from 1 to 13 and the possibility of creating combinations, thanks to color and suit. Coins and dice had been used before (e.g., in Calandri's work), but the use of cards is described for the first time in Paciolo's work.

None of the effects are necessarily executed with cards, but in some instances he says that the same trick done with cards is more deceptive. Another interesting thing is the justification that Paciolo, being a Franciscan Monk, gives to the use of such objects as dice, cards, and "*trionphi*" (tarot cards), usually considered of an unbecoming nature for a religious person: he uses them not to gamble but to demonstrate the power of numbers in an easily comprehensible manner.

Let's see now in which chapters Paciolo describes the use of cards:

- *Quarto effecto. de un numero in tre parti diviso, etcetera* [p. 30][2] (Fourth effect: of a number divided in three parts, etc.). He gives the description of this effect with numbers, dice, and cards.

- *Quinto effecto. de un numero diviso fra 4, o vero in 4 parti* [p. 36] (Fifth effect: of a number divided by four, or in four parts). This is an interesting principle, still in use today. He mentally assigns a number to each of four spectators, each one holding a different card. With the rule given, you know who has which card.

- *XXX effecto. de numero pensato, multiplicato più volte gli suoi producti per diversi o medesimi numeri, trovare l'avenimento partito* [p. 87] (30th effect: of a number thought of, multiplied several times, the products by the same or different numbers, to find the resulting division [less literally] to find a number thought of, from the result of its being multiplied several times, by the same or different numbers). In this chapter, Paciolo describes the use of a confederate, a child, who secretly holds a paper where all the results of the possible multiplications are written; and when the performer asks for a product, the child is instructed how to answer properly. To make this easier for the child, and also more impressive, he suggests putting the child in another room. In this way the child can easily read directly from the paper without being seen. Then Paciolo suggests that the sequence of the tricks should follow a path with high and low, thus obtaining more emotional involvement; these are practically the rules of a theatrical performance. He also says that by instructing the child, it is

---

[2]Page numbers refer to the Peirani Marinoni transcription.

A Lifetime of Puzzles

possible to secretly communicate to him by code words, or gestures, or signals (coughing, tapping with a knife on the table, and so on), or by numbers. He mentions a magician, whose name was Jasonne da Ferrara, who was performing such effects with a boy in gentlemen's houses where Paciolo was personally attending.

- *XXXV effecto. de saper trovare 3 varie cose divise fra tre persone, et 4 divise fra 4, et de quante vorrai* [p. 112] (35th effect: to know how to find three different things distributed among three persons, and four distributed among four, and of as many as you wish). I mention this chapter, although no card tricks are described, because Paciolo suggests memorizing the operations by verses, a mnemonic method that will be widely used later (and still is) to remember the order of the sequence of the cards in a stacked deck.

- *XL Capitolo. de doi cose una per mano divise o ver fra doi, o ver doi numeri inequali, paro et imparo, senza alcuna interrogatione sapere* [p. 118] (Chapter 40: to know of two things, distributed one in each hand or between two people or two unequal numbers, one even and one odd, without any questioning). This chapter gives a method for guessing, between two spectators, who has an odd number and who has an even number of things. In the second example, two cards, one odd and one even, are used. They are thrown on the table face up, so you can see their value.

- *LXIIII Capitolo. d'un numero pensato per via de un cerchio* [p. 151] (Chapter 64: [to know] of a number thought of by means of a circle). In this trick, cards are the perfect things to use. It is the classical trick The Tapping Trick or Tapping the Hours, where a known number of cards are placed in a circle on the table, face down (you only know the value and the order of them, which is progressive). Now a spectator is invited to think of a number not higher than twelve (if you are using twelve cards); then by given instructions, he has to count starting from a point and will finish on a card that when turned over will have the value of the number that was thought of. This trick is also described in the Galasso booklet.

- *LXXX Capitolo. De le gentileza che a le volte si fanno per vie naturali senz'altro calcolo* [p. 177] (Chapter 80: of persons who instantly determine [a number] by natural means without other calculation). I mention this chapter where no card tricks are found but a very interesting principle is explained. Paciolo names two persons who both used this principle: one is Francesco de la Penna, and the other the already-mentioned Giovanni de Jasone da Ferrara. The technique is what we today call in card magic "estimation." The performer knows by experience how many nails, or walnuts, or whatever, can fill a certain container. In a performance, he invites a spectator to fill, for instance, a bottle with nails and predicts how many will fill it, with a very close approximation that amazes the audience. In the same way, he can say how many walnuts can be held in a fist, and so on.

No more card tricks are described in the manuscript, but a lot of magic and amusing physical principles as well as puzzles can be found.

The second part of the manuscript is largely devoted to geometry, but from Chapter LXLIIII [p. 275] on, Paciolo describes a series of puzzles and some hydraulic principles, as well as some optical recreations, some stunts based on physics, secret writings, and a few magic tricks. Many of these are described later in Cardano's and Della Porta's works.

The third part of the manuscript has no card tricks but is filled with very interesting things. Just to name the most intriguing:

- *Capitolo VIII*, second paragraph [p. 334]. Using prepared pieces of paper (some of which float on water, some not), you can make a friend become a victim who will pay a penalty.

- *Capitolo IX* [p. 334]. Paciolo describes the right-to-left mirror writing of Leonardo, which can be read with a mirror.

- *Capitolo X* [p. 335]. How to write a sentence on the petals of a rose or other flower.

- *Capitolo XI* [p. 335]. How to engrave letters on iron by the use of chemicals. This technique will be developed soon afterward for engraving and printing.

- *Capitolo XXIII* [p. 342]. A method for washing the hands in molten lead without being hurt, a stunt used by mountebanks and fireproof performers.

- *Capitolo XXXVI* [p. 348]. How to cut a pigeon's neck with a knife without killing him.

- *Capitolo XXXVII* [p. 349]. How to kill a pigeon by hitting its head with a feather ... this time really killing it!

- *Capitolo XL* [p. 350]. How to make an egg crawl on a table. A leech is placed into a hollowed egg. The hole in the egg is closed with wax. The egg is placed next to a vase filled with water and the leech will feel the water and start to move to reach it, making the egg crawl on the table. A similar method (using a bug) will be found described in many pamphlets of secrets of the sixteenth and seventeenth centuries.

- *Capitolo XLI* [p. 350]. A very "modern" method to make a coin go up and down in a glass filled with water using powder of "calamita" (magnetite).

- *Capitolo XLIIII* [p. 352]. Another effect of a coin dancing into a glass on your command using a woman's hair attached at one end to the coin with wax and the other end, again with wax, attached to your finger (a method still in use).

- *Capitolo XLVI* [p. 353]. How to eat tow and spit fire—a classic, very old trick, still in use today by street performers and pseudofakirs.

- *Capitolo LXVI* [p. 362]. How to cut a glass spiral shaped so that it works as a spring.

- *Capitolo LXXIII* [p. 364]. An optical illusion demonstrating the inability to compare horizontal and vertical distances.

Acknowledgments. My grateful thanks to all the friends who continuously support me and my work; for Atlanta's G4G6, my thanks go especially to Bill Kalush, Mark Setteducati, and P.G. Varola.

Part III

Move It

Move it

Railway Mazes

Roger Penrose

In homage to Martin Gardner for his 90th birthday, I wish to describe a very simple idea that has been familiar to me for many, many years, but, strangely, I have never seen it discussed in any detail. This is the notion of a *railway maze*.[1] As far as my own experience is concerned, the basic idea came from my father (Lionel S. Penrose, FRS, 1898–1972, who had been Professor of Human Genetics at University College, London, from 1945 until his retirement from the university), although this idea is such a simple one that I cannot imagine that it had not been originated long, long ago, in the shadows of antiquity. I was certainly a child when he first acquainted me with the idea, but I have no recollection of how old I was. The maze consists of a connected network of smooth curves drawn in the plane (though planarity plays no critical role here), branching and rejoining at various places. The object is to find a smooth path along the curves, from the starting point S to the final point F. We may think of a railway engine, *with no reverse gear*, travelling along the track from S to F. I recall, from a quite early age, my father showing a fairly simple example, perhaps like the one illustrated in Figure 1, which I found to be surprisingly tricky to do, considering its simplicity. Most routes return the en-

---

[1] I use the English term *railway* here, rather than the American *railroad*, because this is what I have been brought up with.

Figure 1. A simple railway maze. Find a smooth path, following the lines, from S to F.

gine back to S, and it is a matter of "puzzle psychology," in the design of such a maze, to try to guide the solver to take the wrong routes in attempting to progress from S to F.

I soon realized that the puzzle could be solved completely without thinking if one chose to search for the route *backwards*, from F to S, and, in fact, there is only one path, with no choices at all! The oddity of this made an impression on me, because the formulation of the *rules* of the maze is completely symmetrical under interchange of S and F (although this particular example is, of course, quite *a*symmetrical in this respect—curiously reminiscent of the way that the second law of thermodynamics operates in our particular time-asymmetrical universe, within the confines of time-symmetric physical laws[2]). If one wishes to make the puzzle more difficult to solve, one can limit the advantage of this kind of "problem solving by backward reasoning" by incorporating a portion of the maze in the vicinity of F that is aimed at making it difficult to start from F, with most routes that start from F returning to F again. (See Figure 2.)

I shall refer to this latter portion of the maze as the *F-directed* part (where the route is difficult to find starting from F) and the original part, which is hard starting from S and guiding the solver back to S, as the *S-directed* part. However, if the F-directed part and the S-directed part are not kept separate from each other, then there are likely to be many different solutions. To avoid this undesirability, one finds that these two parts have to be connected only by a single line. If a point C on this connecting line is found, then

---

[2]For matters relating to the profound issue of symmetry/asymmetry in physical laws, see Martin Gardner's beautiful book *The New Ambidextrous Universe* [2]. For some further information on these matters, see my book *The Road to Reality: A Complete Guide to the Laws of the Universe* [3], particularly Chapter 27.

A Lifetime of Puzzles

Figure 2. To make it a little more difficult for those who like to solve mazes by "backward reasoning," an *F*-directed part has been added.

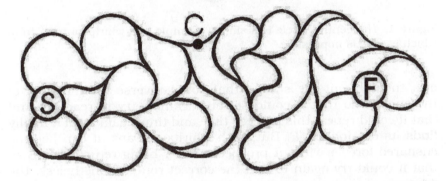

Figure 3. The maze is still readily solved if a point *C* can be located that separates the *S*-directed part from the *F*-directed part, since the routes from *C* to *S* and from *C* to *F* are now trivial.

the maze is extremely easily solved, simply by separately imagining travelling from *C* to *S* and from *C* to *F* and then connecting the two routes. In fact, with merely the ingredients that we have used so far, it seems to be difficult to disguise this central point, and such mazes appear to be very easily solved. (See Figure 3.) Accordingly, some new features need to be incorporated in order to make railway mazes interesting.

I then found that a very useful obscuring device was to introduce "whirlpool" traps that, once entered, would force the engine to circle indefinitely without ever being able to escape. In a sense, these somewhat change the character of the puzzle, as one might imagine that whenever an engine returns to *S*, after having originally set out from *S*, then it might take advantage of the presence

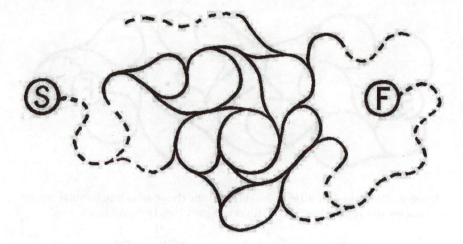

Figure 4. The central circle is a simple example of a whirlpool trap, from which escape is impossible.

of a turntable at the station that could reverse the direction of the engine, so that it could set out again to try to reach F, and that it could repeat this process time and time again until it finally finds its destination F! But, with whirlpool traps, it can become ensnared forever without prospect of ever being returned to S so that it could try again to find the correct route. Nonetheless, the presence of such whirlpools does not seem to violate the intentions of the original prescriptions of a railway maze.

Such whirlpools can be highly localized, as in the circle in the middle of Figure 4, or they can even globally surround the entire maze, as in the example depicted in Figure 5. Also, multiple whirlpools like the example shown in Figure 6 can be involved. Because the full extent of the basin of a whirlpool (that region from which there is no escape but to enter the whirlpool) may not be immediately obvious to the eye, and because this region may connect to both the S-directed and the F-directed parts of the maze, the central point C may not be so easily located, as its removal no longer needs to separate these two parts.

Additional Subtleties

At this point, and before reading further, the reader might care to try the two railway mazes depicted in Figures 7 and 8, which ap-

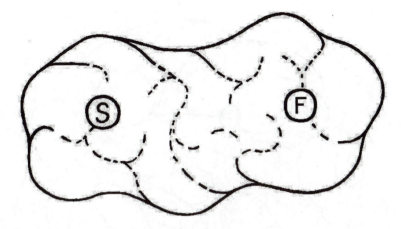

Figure 5. A whirlpool trap need not be localized and can even surround the entire maze.

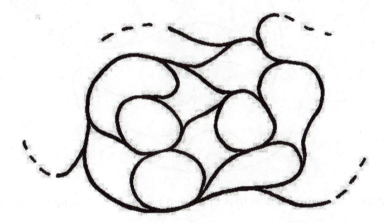

Figure 6. An example of a multiple whirlpool trap.

peared in an article that I wrote jointly with my father in 1958 [4]. (The maze in Figure 8 is somewhat whimsical.) Figure 9 shows another maze that I designed for James Dalgety's "Millennium Bench" by the entrance to St. Mary's Church in Luppitt, Devon, U.K. [1]. There is a particular simple feature that these mazes incorporate that is easy to overlook and, if unappreciated, may lead to the conclusion that solving the maze is impossible! I had an interesting experience of this nature while visiting Princeton University on a NATO Fellowship, in 1959. At the Physics Department's Christ-

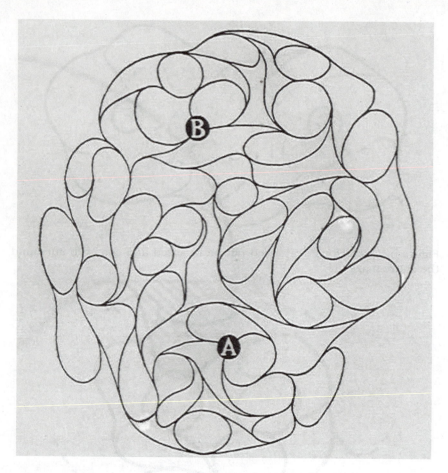

Figure 7. A railway maze that appeared in the *New Scientist* in 1958.

mas party, I put a railway maze on the blackboard (together with some pictures of impossible objects). Various faculty members collected around the board, including the distinguished mathematical physicist Eugene Wigner. After spending some time trying to solve the maze, and learning that I was responsible for it, Wigner turned to me and complained, in his distinctive Hungarian accent, "It is impossible!" He was very surprised when I showed him the solution. I remember my father expressing some delight at this when I later informed him about it by letter, and he referred to the maze as the "stuck-Wigner" puzzle.

The new feature that is being incorporated into these mazes is a track-reversing key loop, whose key role may be easily overlooked

Figure 8. A whimsical example that accompanied that of Figure 7.

by the solver. To use such a device effectively, the maze should be constructed so that, by the time the solver finally encounters the key loop, there will already have been numerous ordinary track-reversal loops that simply guide the engine from S back to S (or from F back to F, for those "solving by backward reasoning"). The point about a track-reversing key loop is that the entire route from S to F must now essentially involve a portion in which the engine reverses its journey along a considerable portion of track. The psychological issue is that the solver may well be fooled into thinking that "this is just another of those tiresome track-reversing loops, sending me back to S, that I must avoid." The key loop has to be

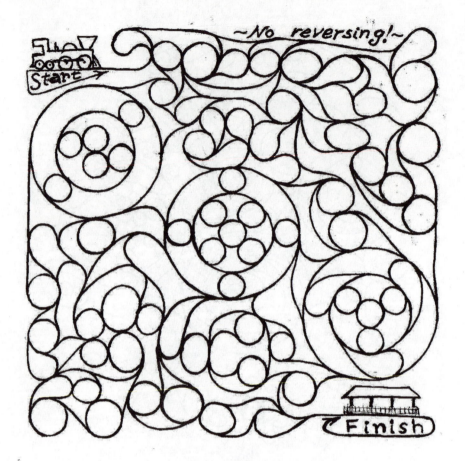

THE RAILWAY MAZE
by Professor Sir Roger Penrose OM FRS
for
THE LUPPITT MILLENNIUM BENCH.
Erected outside St.Mary's Church, Luppitt, Honiton, Devon Ex14 4RZ
© 2000 Luppitt Parish Council
further information http://luppitt@puzzlemuseum.com

Figure 9. A railway maze on the Luppitt Millennium Bench.

well hidden, and there are many ways to hide it, usually involving the proximity of one or more whirlpool basins, but there are other ways of sometimes catching the solver unawares. (See Figures 8 and 9.)

A Lifetime of Puzzles

There is a point of relevance to the "purist" puzzle maker for the construction of railway mazes that incorporate such track-reversing key loops. The solution to a maze containing such features is, technically speaking, always *non-unique*. For the "loop" part of the solution path can always be traversed in either of two opposite directions. Of course, the two solutions arising in this way are not "very" different from each other (and are always of the same length), but the purist might worry! My own attitude to this has been simply to make the reversal loop "small," so that it is easy to regard this ambiguity as unimportant. There is a more serious type of "cook" (i.e., unintended solution) that can easily arise if one is not careful to make sure that the stretch of track that is traversed in both directions is separate from both the S-directed and the F-directed region. (See Figure 10(a–b).) One can also incorporate *two* (or more) track-reversing key loops into a railway maze. In this case, to avoid serious cooks, one must make sure that no part of the track that joins the key loops lies in either the S-directed or F-directed region. However, this situation will always result in paths from S to F that might be regarded, technically, as cooks. For, after employing the first key loop, one can use the second to return to the first and then back to the second again before moving on to F; see Figure 10(c). This is clearly an inefficient procedure, resulting in a route of excessive length, and I think that it is fair to ignore this kind of cook. This sort of thing also occurs with the situation of Figure 10(d), where a "non-key" reversal is incorporated into the route out from the key, seeming to have no value except to introduce this kind of technical cook. In fact, such things can have a psychological value in helping to disguise the intended solution to the maze.

Toy Track Railway Mazes

Let us now move on to another (but related) type of railway-track puzzle. Having had a four-year-old child with a passion for toy train tracks (at the time of writing, he was five)—where I must confess I get great enjoyment joining in the fun and constructing complicated track designs of my own—I have found that there is a quite different kind of puzzle that the idea of a railway maze provides. Here, one envisages that there is an unlimited supply of straight and curved tracks, where bridges and crossings can, if desired, also be supplied in unlimited numbers. In addition to these, one is provided with a definite number (in fact, an even number) of

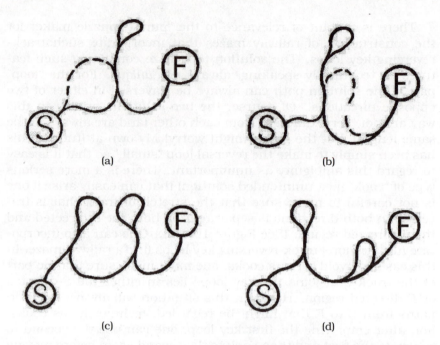

(a)   (b)

(c)   (d)

Figure 10. Track-reversing key loops. (a) The part of the path that needs to be traversed in both directions must be separate from the S-directed part, or (broken curve) a serious cook arises. (b) It must also be separate from the F-directed part if a serious cook is to be avoided. (c) Two track-reversing key loops are legitimate, although technically a cook arises from the possibility of doubling back and forth between the loops. (d) The same could hold even for non-key loops, if directed inwards as shown.

railway points. Each of these enables the bringing together of two tracks to produce one or, conversely, the dividing of a single track into two. I shall refer to the end that joins on to a single track (the track that is being made to bifurcate by the point) as the *single end* and to each of the other two (together constituting the bifurcation itself) as the *double ends*. (See Figure 11.)

Again, we are to have an engine with no reverse gear (or a reverse gear that is not supposed to be used), where the engine simply continues indefinitely in its (local) forward direction. The railway points are of the type that each has a switch on it that can be set in one of two possible states ("settings"), so that if the engine enters along the single end, then this setting determines which of the two double ends it exits along. If, on the other hand, the engine enters along either one of the double ends, then it simply

Figure 11. Railway points with a spring. There are two possible settings for the point. For the setting shown, an engine entering along the single end P must exit along the particular double end R; if the engine enters along either double end Q or R, it will exit along P, leaving the setting undisturbed.

leaves along the single end without altering the setting. (The Lego Duplo^{TM} train-track points have this character—and also some special Brio points—having a switch attached, for fixing the setting, where there is a spring on the component determining the setting, so that if the engine enters along the "wrong" double end, then it simply pushes this component aside, the spring ensuring that the component returns to its original position, as predetermined by the setting.)

The object of the puzzle is now, for a given (even) number $2n$ of railway-point pieces, to construct a train track, with no unattached ends, in such a way that, for the appropriate collection of settings, the train will continue indefinitely, without omitting any stretch of track. We find that, for any $n > 0$, such tracks can be constructed. In fact, for any given $n > 1$, there are likely to be *several* solutions for the collection of settings, so as a refinement we demand that the entire track arrangement must be such that there is essentially a *unique* collection of settings of all the points having the result that no section of track is omitted in the course of the engine's journey.

The word "essentially" is needed here, owing to the fact that, given a particular track arrangement, it is *always* the case that, for any collection of track settings that gives a solution, we also get a solution if all the settings are *reversed*. (Here, "solution" refers to a collection of settings for which the engine will necessarily visit all stretches of track during its journey.) Why is this? Imagine that, upon reversal of all point settings, we also reverse the direc-

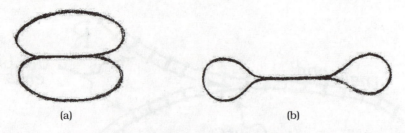

(a)   (b)

Figure 12. A closed track will have $2n$ points. Here, the two possibilities for $n = 1$ are shown, for which there are no whirlpool subregions. (a) In this case there is no reversal, so there can be no solution for the settings in which the engine must visit all track sections. (b) In this case, every setting gives a solution, but not an essentially unique one.

tion of the engine. Consider one junction, as depicted in Figure 11. Suppose that the engine enters along the single end P, and the setting forces it to exit along the particular double end R. Because the track portion entering along the remaining double end Q must also be traversed during the course of the engine's journey, this must be when the train is moving in the opposite direction, namely inward toward the junction, and it must leave at P along the single end, following the opposite direction from which it was first considered to have entered the junction. If we now reverse the setting, together with all the engine directions just considered, then we get a situation, at the junction, that exactly mirrors what has been just described, but with Q and R interchanged. The same will apply at all the other junctions. Accordingly, we always have the necessary ambiguity that a reversal of all the settings, all at once over the entire network, will transform one solution into another. So, it is reasonable to regard two solutions as being essentially the same if one is obtained from the other simply by the reversal of the settings of *all* points.

The case $n = 0$ is clearly trivial, because the track simply consists of a single loop. If we take $n = 1$, there are, topologically,[3] only two possibilities that do not contain proper whirlpool subregions (the presence of which would obviously forbid any solution), namely those shown in Figure 12(a–b). That of Figure 12(a) does not allow for any reversing of the engine, so there can be no collection of settings that provides a solution. The arrangement of

---

[3]This is topology in the sense of *differential* topology, so we are concerned with *smoothness*, and not just the ordinary topological notion of continuous deformation (which need not be a smooth deformation).

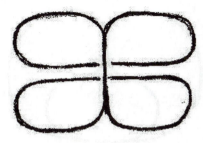

Figure 13. The same as Figure 12(b), but giving a more interesting track arrangement (with a bridge).

Figure 12(b) gives a solution for every collection of settings, but there are four of these, so the solution is not essentially unique. It may be remarked that although this arrangement is very simple as drawn in Figure 12(b), this simple topological organization can be disguised in many ways, and many quite interesting actual track arrangements can be constructed, such as that shown in Figure 13.

For every larger value of n (i.e., for $n \geq 2$), we find track configurations for which the collection of settings is essentially unique. For example, when $n = 2$, the arrangement of Figure 14 will achieve this. This configuration can be rearranged into the symmetrical configuration in Figure 15(a), which takes advantage of a bridge, and the essential uniqueness is not hard to ascertain using the symmetry. This construction easily generalizes to any even $n > 2$, giving essential uniqueness, the $n = 4$ case being illustrated in Figure 15(b) (where I have used a multiple flyover at the center!). For *odd* $n > 2$, we can use the simple-looking configurations of Figure 16(a–b), giving essential uniqueness, where I have explicitly illustrated the cases $n = 3$ and $n = 5$. In fact, all of these essentially unique configurations can, with a little rearrangement, be expressed uniformly as part of the same series, shown in Figure 17. There are many other essentially unique configurations, such as that depicted in Figure 18 for $n = 4$, but of course in most cases where a solution exists, it is not essentially unique, such as the example in Figure 19 for $n = 6$. It's a nice exercise to find all the solutions in this case (there are just three essentially different ones) and to prove essential uniqueness in the other examples.

Martin Gardner would have written a great article on railway mazes in *Scientific American*. Perhaps he did, and I missed it!

Figure 14. For *n* = 2 there is an essentially unique system of point settings that provides a solution (i.e., so that the engine is compelled to visit all track segments).

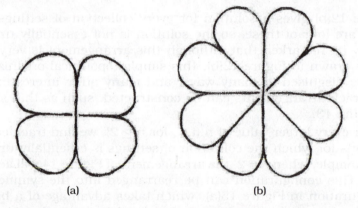

(a)   (b)

Figure 15. For even *n* we get solutions for essentially unique point settings with these configurations: (a) the case *n* = 2 (equivalent to Figure 14), (b) the case *n* = 4.

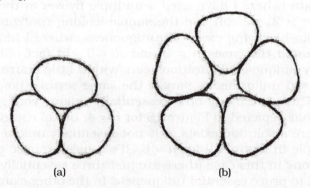

(a)   (b)

Figure 16. Configurations giving solutions with essentially unique point settings for odd *n* > 1: (a) the case *n* = 3, (b) the case *n* = 5.

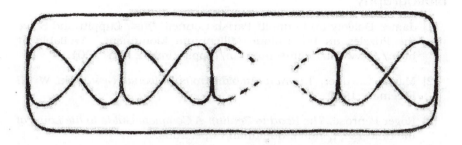

Figure 17. The configurations of Figures 15 and 16 redrawn so that it can be seen that they are parts of the same sequence, for all $n > 1$.

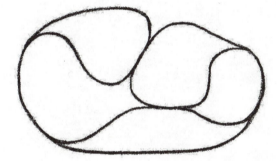

Figure 18. There are also other configurations giving essentially unique solutions; here, $n = 4$.

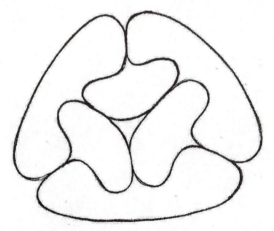

Figure 19. A non-essentially-unique case for $n = 6$; here, there are three essentially different collections of point settings.

Bibliography

[1] James Dalgety and Luppitt Parish Council. "Visit Luppitt and Solve the Puzzles on the Unique Millennium Monument." Available at http://www.puzzlemuseum.com/luppitt/lmb02.htm, 2001.

[2] Martin Gardner. *The New Ambidextrous Universe*. New York: W. H. Freeman, 1990.

[3] Roger Penrose. *The Road to Reality: A Complete Guide to the Laws of the Universe*. London: Jonathan Cape, 2004.

[4] L. S. Penrose and R. Penrose. "Puzzles for Christmas." *New Scientist* (December 25, 1958), 1580–1581, 1597.

Mechanical Mazes

M. Oskar van Deventer

Most people know mazes as printed pencil-and-paper problems, where one has to draw a path from the start to the finish, such as Martin Gardner describes in Chapter 10 of *The Second Scientific American Book of Mathematical Puzzles and Diversions* [1]. Colorful plastic mazes, where the goal is to remove a ball, are a direct translation from pencil-and-paper mazes into a mechanism. The same is true for life-size mazes made from hedges, mirrors, or corn, where the goal is to find the exit. This article is *not* about these types of mazes. Rather, this article is about mechanical puzzles, which may not look like a maze at all, but for which the solving experience feels like solving a maze. Let me explain with some examples.

Cooksey Cylinder

My fascination with mechanical mazes began in July 1983, when I visited Edward Hordern for the first time. One of the puzzles he showed me was the *Cooksey Cylinder*, a red cylinder surrounded by a bronze pipe with a pattern of slots. (See Figure 1.) A small pin follows the slots. What made this puzzle so special is that the pin could be pushed to the other side. This way, the actual maze is distributed over two sides of the cylinder, which makes it quite hard to solve. One way to solve the maze is by mapping it out,

Figure 1. *Cooksey Cylinder*, © Mr. Cooksey, prototyped by Pentangle.

Figure 2. Map of the *Cooksey Cylinder*.

noting down rotations as horizontal moves and shifts as vertical moves. To my large surprise, the resulting map is quite simple. (See Figure 2.)

A Lifetime of Puzzles

Key Maze

In 1987, I designed *Key Maze*. (See Figure 3.) The disk can be turned and moved up and down along the shaft of the key. The notches and pins of the key interact with those in the disk, making this puzzle a maze. The maze can be mapped in a similar way as the *Cooksey Cylinder*, by unrolling the cylindrical surface. I based the puzzle on a maze-like fractal called the Dragon Curve. You may recognize the fractal-ness of the maze in its map. (See Figure 4.) The map also makes clear why people get lost in this maze. The maze has several loops. Moreover, you have to turn the key a full 720 degrees to solve the puzzle. If you move down too fast, you'll get stuck in the dungeons of this maze.

Figure 3. *Key Maze*, produced by Bits & Pieces, http://www.bitsand pieces.com.

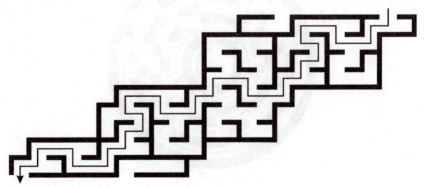

Figure 4. Map of *Key Maze*.

Möby Maze

Möby Maze is a mechanical maze that I designed in 2003. It was produced by rapid prototyping, as that was the only way to build such a twisted shape. (See Figure 5.) The movement of the red ring is obstructed by obstacles at "both" sides of the Möbius ring. It is not possible to map the Möbius surface in a plane, because of its half twist. The most effective way to map this maze is by drawing it twice. (See Figure 6.) A 360-degree tour on this map corresponds

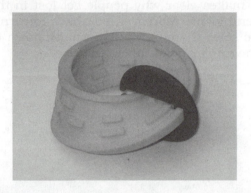

Figure 5. *Möby Maze*, produced by George Miller, http://www.puzzle palace.com.

Figure 6. Map of *Möby Maze*.

to a 720-degree tour on the Möbius ring. You may check for yourself that the upper half of the map is essentially identical to the lower half. The maze is quite simple, but still people will get lost. At some point, there is a sharp Y-turn to the exit. People have the tendency to preserve momentum and may go round and round in the long loop around the center many times.

Tube Maze

Tube Maze is a plumbing puzzle that I designed in 2003. It too is rapid prototyped. The object is to get the ring off the tube structure. At every T intersection, the ring can move straight or make a turn. In its starting state, the asymmetric flag blocks the ring. (See Figure 7.) The ring needs to be turned upside-down to be able to take it off. The puzzle is essentially a parity problem. (See Figure 8.) At one section of tube, the parity of the ring is changed, where the thick circuit transitions into the thin circuit. One should pass this section exactly once (or three times, or five) to solve the puzzle. The map is a rather abstract representation of the actual puzzle, and one needs an additional table to match the map intersections with the T intersections of the maze.

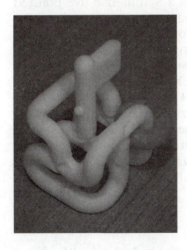

Figure 7. *Tube Maze*, produced by George Miller, http://www.puzzlepalace .com.

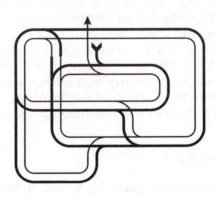

Figure 8. Map of *Tube Maze*.

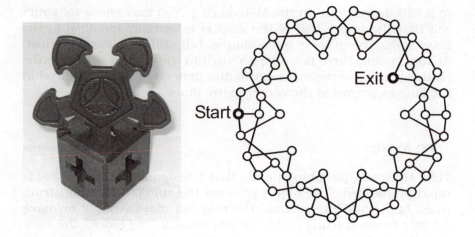

Figure 9. *O'Gear*,
produced by Hanyama,
http://www.hanayamatoys.co.jp.

Figure 10. Map of *O'Gear*, mapped out by
Peter Winkler.

O'Gear

In 1995, I designed *O'Gear*, which won an international puzzle
design competition award in 2001. It is a symmetric five-pointed
star that moves around a cube. (See Figure 9.) Some of the cube
edges are sharp, preventing the star from passing. The object is
to remove the star from a given starting position. The star can
be in ten orientations at each of the surfaces of the cube: five
orientations facing forward and five backward. As a cube has six
faces, the puzzle has a total of $6 \times 10 = 60$ states. Peter Winkler
found a way to map the state diagram such that the symmetry
of the puzzle is reflected in the symmetry of the map: five-fold
rotational symmetry and mirror symmetry. The map gives a good
indication of the complexity of the puzzle, although people may
find it hard to use this diagram to actually solve the puzzle.

Big Wheel

In 2001, I designed *Big Wheel*, in which the red wheel with eleven
teeth runs like a gear on the black floor piece. (See Figure 11.) The
object of this puzzle is to take the wheel off the floor. The floor
piece has five round holes, and the wheel has five "special" teeth.
When a special tooth matches up with a round hole, the wheel can

A Lifetime of Puzzles

Figure 11. *Big Wheel*, produced by Bits & Pieces, http://www.bitsandpieces.com.

Figure 12. Maps of *Big Wheel*: (a) 22-state map and (b) 25-state map.

be turned 180 degrees horizontally. The wheel can roll out when the exit tooth matches up with the exit hole. Mapping this maze is ambiguous, as the "states" of the maze can be defined in more than one way. One mapping is by looking at the orientation of the wheel at a particular position on the floor. As the gear has eleven teeth and two sides, the gear has $11 \times 2 = 22$ states, resulting in a 22-state map. Andrea Gilbert made a different analysis. As there are five holes and five special teeth, she defined the $5 \times 5 = 25$ combinations as states of the maze, resulting in a 25-state map. As you can see from Figure 12, the two maps are significantly different. However, the topology of both graphs is the same, as they both describe one and the same maze.

GGG or 1.5 Horses

I designed *GGG* in 1995, and a simplified version of it won an international puzzle design competition award in 2002. The puzzle has three G-shaped rings that are linked together. (See Figure 13.) The object is of course to take them apart. Most puzzle collectors would probably categorize this puzzle as a disentanglement puzzle. However, the puzzle was designed as a maze and can be analyzed as a maze. The different ways that the pieces can be tangled can be defined as "state" of the maze. It was calculated that this puzzle has $(12 \times 2) + (8 \times 2) + 26 + 2 = 68$ states. Needless to say, this puzzle is very difficult. Figure 14 shows the resulting state map, which is explained in more detail in [1]. By the way, the name *GGG*

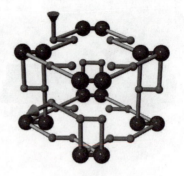

Figure 13. *GGG or 1.5 Horses*, produced by James Dalgety, http://www.puzzlemuseum.com.

Figure 14. Map of *GGG or 1.5 Horses*.

or 1.5 Horses is a pun by James Dalgety. Apparently, the British refer to a horse as a "gee-gee" to their children. As the puzzle has three G-shaped pieces (gee-gee-gee), it is one-and-a-half horses.

Epilogue: The Cooksey Cylinder

The puzzle it all began with, the *Cooksey Cylinder*, remains obscure. Nine prototypes were made by Pentangle, but they decided not to pursue this puzzle because of the production cost. Mr. Cooksey, the inventor of this fascinating puzzle, has vanished without a trace. Thanks to a trade with James Dalgety, I am now the proud owner of prototype #4 from the Hordern–Dalgety collection. I hope that this classic design will be rediscovered by a puzzle producer sometime in the future, so that more people will be able to enjoy it in its physical form.

Bibliography

[1] Martin Gardner. *The Second Scientific American Book of Mathematical Puzzles and Diversions*. Chicago: University of Chicago Press, 1987.

[2] Oskar van Deventer. "GGG or 1.5 Horses." *Cubism For Fun* 61 (2003), 23–25.

A Lifetime of Puzzles

Insanity Puzzles: Instant and On-the-Spot

Rik van Grol

Until recently I thought that there had been only three major puzzle crazes: the *Fifteen Puzzle*, the *Tangram*, and the *Rubik's Cube*. Now I know better. In 1967, a puzzle was brought to the market that sold over 12 million copies in no time: *Instant Insanity*. I have had a similar puzzle for a long time, but I never knew it had been so popular. I always loved this puzzle and designed a derived puzzle for the International Puzzle Party in 2003. Wanting to present the puzzle and its solution at G4G6, I started investigating the background of insanity puzzles and found a mountain of material in terms of both books and Internet sites. In this paper I first present my version of the story on *Instant Insanity* and reiterate the solution technique for this puzzle (a method using graphs). Then, I introduce my newly designed puzzle called *On-the-Spot Insanity* and present its solution. Finally, I discuss the possibility of using graphs to solve this puzzle.

Martin Gardner wrote about the puzzle twice: in *Fractal Music, Hypercards and More Mathematical Recreations from Scientific American Magazine* (W. H. Freeman, 1992, Chapter 6), and in *Mathematical Magic Show* (MAA, 1989, Chapter 16).

Instant Insanity: The Concept

Instant Insanity consists of four colored cubes that need to be lined up in such a way that each of the four (long) sides shows four different colors; see Figure 1. The four cubes are colored in four colors only, which implies that some colors on each cube occur more than once. Solved, the order of the cubes is irrelevant, but finding the correct orientation of each cube is what this puzzle is all about. As each cube has 24 possible orientations, solving the puzzle by trial and error will be a time-consuming task. In other versions of this type of puzzle, with five or even six cubes that need to be lined up, the method of trial and error is even less fruitful. *Instant Insanity* is the name by which this puzzle is best known, but it is known by several other names, and in different forms, as I will show in the next section.

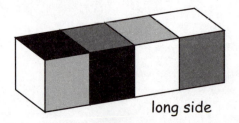

Figure 1. Instant Insanity.

"Instant" History

Instant Insanity is a puzzle designed by Franz O. Armbruster. He licensed it to Parker Brothers, who brought it to the market. They sold over 12 million copies in 1966–1967. For the colors and the exact configuration in which Instant Insanity was sold, I refer to the Insanity puzzles overview in Table 1 below. The puzzle, as Armbruster designed it, was not the first version of this puzzle. According to T. H. O'Beirne in 1965 [1, pp. 112–129], earlier versions exist that were over fifty years old at that time. Jerry Slocum and Jack Botermans [5, p. 38] mention a patent by Frederick A. Schossow from Detroit from around 1900, but their only source for this is O'Beirne. David Singmaster [3, 4] however reports Schossow's patent and several others:

- Schossow's Patent (1900) for four cubes;

- Moffat's Patent (1900) for up to six cubes;

- Meek's UK Patent (1909) for four cubes;

- Wyatt's Puzzles in Wood (1928) gives a six-cube version considering all six directions.

O'Beirne, who actually presents the most extensive description and analysis of the *Instant Insanity*-type of puzzle, is not very revealing where it concerns the origins and the inventors. However, he does provide a good listing of this type of puzzle available in 1965 prior to the appearance of *Instant Insanity*. Jerry Slocum's book nicely complements this by providing the illustrations. The puzzle Schossow designed was brought to the market with the name *Katzenjammer Puzzle*. A German book by Rüdiger Thiele explains, "This puzzle is known for a long time. First it made itself well-known in the English-speaking part of the world with the name 'Katzenjammer.' Later it came back to us with the name 'Instant Insanity'" [6, pp. 75–77].

Instant Insanity uses four cubes with colored sides. The original *Katzenjammer Puzzle* used the four card suit symbols. During the First World War, the puzzle was brought to the market with the flags of the allied nations. Jerry Slocum shows two versions, one with four cubes and one with five. O'Beirne reports only a five-cube version, called *Flags of the Allies*. He reports the flags being those of the following nations: Belgium, France, Japan, Russia, and the United Kingdom. O'Beirne furthermore reports a version showing groceries; see Table 1. He also reports a recently circulating (1965) version showing a red triangle, a blue triangle, a bottle, and a glass. (He suggests that English readers will need no further clue to its source—I have no idea what it means.) O'Beirne mentions that the puzzle used to be called *Tantalizer* and even *The Great Tantalizer* (a version with colored faces). He finally mentions a version with colored dots. O'Beirne continues by telling the reader that the puzzles he found appeared to be different but are abstractly identical. The different colors and pictures have complete correspondence, as can be seen in the top half of Table 1. I share O'Beirne's amazement that people or firms bringing out a new version of a puzzle do not make the effort to design their own version, but simply replace the original pictures with something new. In duplicating an idea, they sometimes make mistakes that

	A	B	C	D
Suit symbols (*Katzenjammer*)	Club	Spade	Heart	Diamond
Groceries 1	Soup	Gravy	Table cream	Custard powder
Groceries 2	Soup	Gravy	Table Cream	Jelly Crystals
Colors (*Instant Insanity 1*)	Red	White	Green	Blue
Colors (*Instant Insanity 2*)	Red	Yellow	Green	Blue

Table 1. Same puzzle, different faces.

result in slight variations. Note that each cube can be replaced by its mirror image without changing the solution (only four sides of each cube are used).

Since O'Beirne's book in 1965, other versions have been brought to the market—to start with, *Instant Insanity*, as explained earlier. *Instant Insanity* is itself a copy of the earlier versions; see the bottom half of Table 1. Remarkably, O'Beirne mentions a correspondent—Mr. C. H. Parker—who told him that he had found a four-cube puzzle with the numerals 3, 4, 5, and 8. The object is to line up the cubes in such a way that on all four long sides the numerals add up to 20. This version is the *Tantalizer* in disguise. The fact that 20 can also be made by 4 + 4 + 4 + 8 and 5 + 5 + 5 + 5 does not matter, because these numerals are not available in high enough frequency on the cubes. It is not clear (but perhaps likely?) that Mr. C. H. Parker is related to Parker Brothers.

Personally, I have three other variations in my possession. The first variant is a German puzzle called *Mutando*, brought onto the market by Ingo Uhl. This puzzle is actually different from *Instant Insanity*. The puzzle has some other goals as well. The first objective is to produce a $2 \times 2 \times 1$ shape in which each side is colored homogeneously. The second objective is the same as for *Instant Insanity*. The second variant is a five-cube version given to my wife in the late 1980s by Kevin Holmes. This five-cube version is two puzzles in one, as it shows the numerals 1 to 5 in five different colors (silver, gold, red, blue, and green). The coloring is independent of the numerals, which allows you to first solve the puzzle in numerals (two solutions) and then in colors (two solutions). The third is another five-cube version that I bought in Japan, to be referred to as *Hanayama* (all other text on the package is in Japanese; the package shows Nob Yoshigahara's logo).

Insanity Puzzles Overview

This section presents the precise configurations of the different insanity puzzles introduced in the previous section.

Katzenjammer Puzzle

Invented ca. 1900 by Frederick A. Schossow, the *Katzenjammer Puzzle* was made of 2.1-cm wooden cubes, printed with the four card suits.

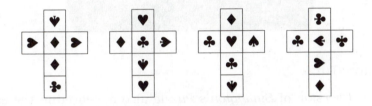

Simington's Puzzle (1)

Appearing sometime in the early twentieth century, the first version of *Simington's Puzzle* was made of 2.7-cm wooden cubes, covered with printed paper showcasing various Simington's products.

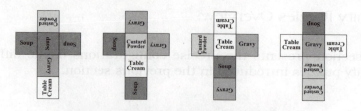

Simington's Puzzle (2)

A second version of *Simington's Puzzle* also appeared in the early twentieth century, made of 2.7-cm wooden cubes and covered with printed paper.

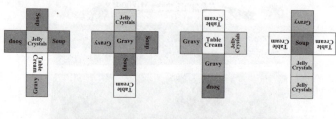

A Lifetime of Puzzles

Instant Insanity (1)

Copyright 1967, 1986, by Parker Brothers Division, this version of *Instant Insanity* was made of 3.3-cm hollow plastic cubes, color-plated in the four colors white, green, red, and blue (in the figure, white, light grey, medium grey, and black, respectively).

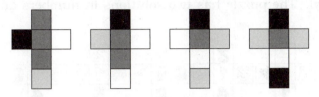

Instant Insanity (2)

Copyright 2000 by Hasbro, this version of *Instant Insanity* is equal to the earlier version above, except for the colors: blue changed to purple and white changed to yellow.

Mutando

Designed by E. Künzell in 1997 and copyright by Ingo Uhl in 2000, *Mutando* was made of 1.8-cm solid plastic cubes, in the four colors yellow, green, red, and blue (in the figure, white, light grey, medium grey, and black, respectively).

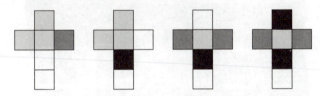

Hanayama

First appearing in the 1990s, *Hanayama* is of Japanese origin and is made of 2.5-cm hollow plastic cubes, marked with colored symbols: purple square, green X, orange circle, yellow cross, and blue four-pointed star. It comes with a display-tray.

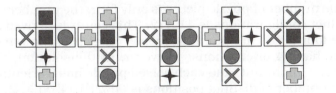

The Trench Tantalizer

Copyright by Kevin Holmes, *The Trench Tantalizer* that appeared in the 1980s was made of 1.8-cm wooden cubes with imprinted colored numbers, 1–5 in white, yellow, green, red, or blue (in the figure, white, light grey, medium grey, dark grey, and black, respectively). The puzzle has two solutions in numbers and two in colors.

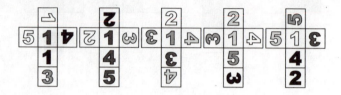

Allies Flag Puzzle

Appearing during the First World War, the *Allies Flag Puzzle* was made of wooden cubes covered with printed paper, the two colors red and blue printed on white.

Solving Instant Insanity and the Like

Solving *Instant Insanity* by trial and error is a tedious process. Although the number of puzzle pieces is only four, the number of possible distinct configurations is 41,472. This count is derived as follows. The four cubes are put in sequence (in 4! possible ways) and each cube has 24 orientations. However, in the solution, the sequence is not important, and each solved puzzle has 8 orientations. Thus, the number of distinct positions is $(4! \cdot 24^4)/(4! \cdot 8) = 41,472$. It

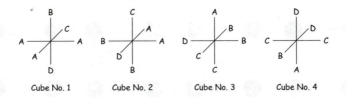

| Cube No. 1 | Cube No. 2 | Cube No. 3 | Cube No. 4 |

Figure 2. Solution to *Instant Insanity*.

is difficult to systematically go through all possible positions without making a mistake.

Before showing a much easier and systematic method, it is important to be able to uniquely identify each of the four blocks of the puzzle, and to name the faces uniquely. O'Beirne describes a straightforward method for this in four steps.

1. Search for the cube with three identical faces; call this cube No. 1, and call these faces the *A* faces.

2. Search for the only cube with two *A* faces; call this cube No. 2. This cube has another pair of identical faces; call these the *B* faces.

3. Search for a cube with one *A* face and two *B* faces; call it cube No. 3. This cube has another pair of identical faces; call these the *C* faces.

4. Call the fourth cube No. 4. This cube has two pairs of identical faces: *C* faces and *D* faces.

Having identified each of the cubes, the solution to *Instant Insanity* can now easily be shown, as in Figure 2.

Avoiding the trial-and-error method and not having the above solution available, there is a very elegant method available to solve Instant Insanity using graphs. According to David Singmaster [4], this graphical method is due to Carteblanche (1947).

The graph method starts by describing the puzzle in the form of a graph. In the graph, the nodes represent the four ways the faces appear, and each link represents a pair of opposing faces on a cube. As each cube has three pairs of opposing faces (or *face pairs*), each cube is represented in the graph by three links. Figure 3 shows the graph representing *Instant Insanity*. If a face pair has two identical faces, the link points back to its origin.

To find the solution shown in Figure 2, the challenge is to find two loops that visit each of the four nodes, using each link at most

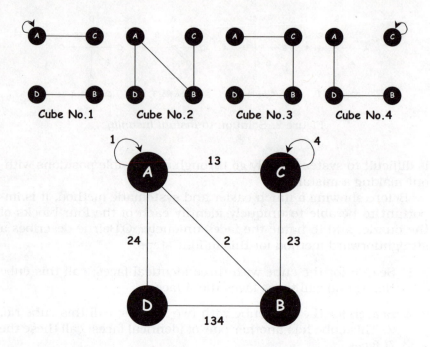

Figure 3. Graph description of *Instant Insanity*.

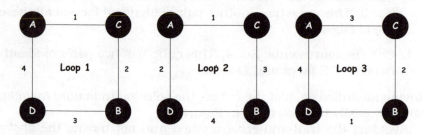

Figure 4. Three possible loops passing all four faces.

once. One such loop basically represents two sequences of all four faces on the opposing sides of the row of four cubes. The number of possibilities in the graph is very limited. Although three individual loops can be found, only Loop 2 and Loop 3 together use each link at most once; see Figure 4.

Although this is the case in *Instant Insanity*, finding two loops is not necessarily the only way to find a solution. The correct requirement is to find two *sets* of loops, each of which includes all

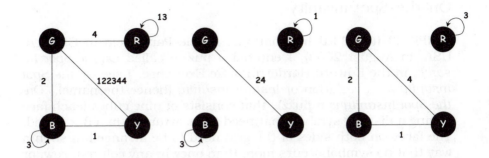

Figure 5. *Mutando* in graph representation (left) and the two sets of loops that form the solution (middle and right).

faces and which together use each link at most once. This is nicely illustrated by the $4 \times 1 \times 1$ solution for *Mutando* shown in Figure 5. This figure shows both the graphical representation of the puzzle on the left, as well as the two sets of loops that form the solution (middle and right).

For both *Instant Insanity* and *Mutando*, using the graph method makes solving the puzzle childishly simple. This is somewhat different when the puzzle is extended by only one cube. Take, for example, the five-cube version *Hanayama*. The graphic representation is shown in Figure 6 (O = Orange, G = Green, P = Purple, Y = Yellow, and B = Blue).

Finding sets of loops in this graph is quite a bit more difficult than in the case of *Instant Insanity*. Then again, the number of distinct positions has grown from 41,472 to 995,328.

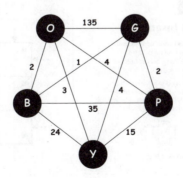

Figure 6. Graphic representation of *Hanayama*.

On-the-Spot Insanity

At IPP 23 (the 23rd International Puzzle Party held in Chicago, USA, in August 2003), I entered a puzzle called *On-the-Spot Insanity* in the Edward Hordern Puzzle Exchange. I see *On-the-Spot Insanity* as a variation of *Instant Insanity* (hence the name). *On-the-Spot Insanity* is a puzzle that consists of nine cubes (each face having a distinct symbol) that need to be arranged in a 3 × 3 grid. The faces on both sides of the grid need to be arranged in such a way that no symbol occurs more than once in any column, row, or diagonal (short or long). A picture of *On-the-Spot Insanity* is shown in Figure 7. The instructions that accompany the puzzle are shown in Figure 8.

On-the-Spot Insanity is a variation of a puzzle called *Osiris* that I found a few years ago. The object of *Osiris* is the same, but it does not use cubes. Each of the nine puzzle pieces has only two sides, which makes the puzzle much easier. A picture of *Osiris* is shown in Figure 9.

Figure 7. *On-the-Spot Insanity.*

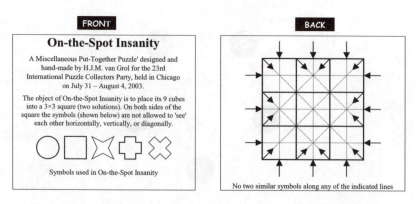

Figure 8. The card with instructions for *On the Spot Insanity.*

Figure 9. *Osiris.*

Solutions to On-the-Spot Insanity

On-the-Spot Insanity has multiple solutions, but I have not yet determined the exact amount. One solution is shown in Figure 10, which thereby also shows the exact configuration of the puzzle. If you want to experience the level of difficulty of solving this puzzle, this is the moment to stop reading. Make the cubes indicated in the figure, mix them up, and try solving the puzzle.

Solving On-the-Spot Insanity

Before discussing an approach for solving *On-the-Spot Insanity*, I will first look at some characteristics of the puzzle. First of all, let us investigate the objective a bit closer. What does it mean that a face may not occur twice in any row, column, or diagonal (short or long)? In a 3 × 3 grid, there are only three distinct kinds of places. Figure 11 shows, for each of these three locations marked by a cross, where a duplicate face may be located, such that the cross cannot "see" it. The left and middle graphics in Figure 11 show that, in each case, there are two locations that cannot be "seen" by the cross. The cross in the third location can see all other locations.

The above leads to the following conclusions:

- Each symbol can at most occur twice;

- Given nine locations, exactly five different symbols are needed to fill the grid (no fewer);

- With five symbols, there are only two possible compositions on the grid; see Figure 12.

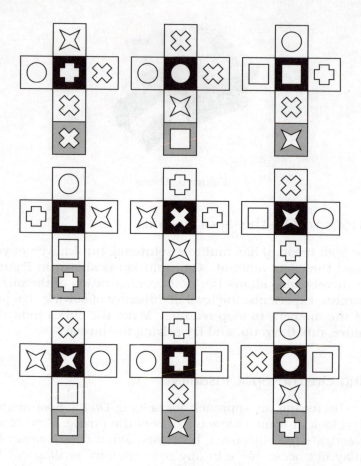

Figure 10. Solution to *On-the-Spot Insanity*.

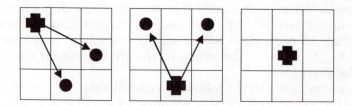

Figure 11. Possible locations for an identical cube face.

Now let us compare the objectives of *On-the-Spot Insanity* and *Instant Insanity*. Although, strictly speaking, the objectives are different, there is a strong similarity. In both cases, the object

Figure 12. The only two possible configurations.

Figure 13. Two sequences X and Z with non-matching cube faces.

is to find sets of non-matching cube faces. The difference is, of course, that in *On-the-Spot Insanity* identical faces may be on the same side of the puzzle, but only when a strict rule is obeyed: they may not "see" each other. Thinking about it a bit longer, you might recognize two sequences of non-matching cube faces: sequence X of length 5, and sequence Z of length 4; see Figure 13. These two sequences are not independent, as they are in *Instant Insanity* (there are only two possible compositions, as shown in Figure 12), but the sequences do suggest that the graph method might also be used here.

Let us investigate the graph method by applying it to *Osiris*. The graph for *Osiris* is shown in Figure 14. Because Osiris has nine puzzle pieces, each with one face pair, there are only nine links. Puzzle pieces 6 and 9 are identical, as are 4 and 7, and as are 5 and 8. They are numbered according to their position in the solution; see Figure 15.

Figure 15 shows two loop sets: Loop set 1 and Loop set 2. It is clear that *Osiris* has only one solution. No other two loop sets can be found in Figure 14. Note that, with the loop sets identified, the puzzle has not yet been solved. The orientation of each puzzle piece is clear (which face should be up), but both sequences still need to be placed on the grid; see Figure 13. Remembering the two possible configurations in Figure 12, we can conclude that the order of the two sequences needs to be the same.

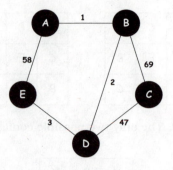

Figure 14. The graph for *Osiris*

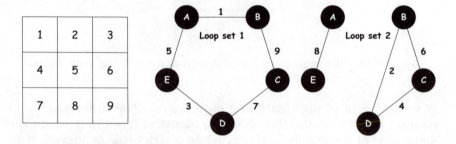

Figure 15. The solution to *Osiris* and its two loop sets

The graph method seems a promising method also for finding a solution to this type of puzzle. With that thought, I started using it on *On-the-Spot Insanity* as well. Figure 16 shows the graph for *On-the-Spot Insanity*, and Figure 17 shows the two loop sets that represent the solution.

Using the graph method seems straightforward, and therefore it is a useful method to solve these types of puzzles. However, the real test, of course, would be to take the graph for *On-the-Spot Insanity* in Figure 16 and to determine the loop sets with which the puzzle could be solved. This is more difficult than I hoped. Finding the two loop sets from Figure 17 is far from trivial. On the other hand, the graph method does facilitate a systematic search. Without this approach, the puzzle is virtually impossible to solve manually. Recall that *Instant Insanity* has 41,472 distinct positions. *On-the-Spot Insanity* has no less than 228,562,145,280 distinct positions. Nine cubes can be placed on the grid in 9! ways. As each of the nine cubes has six distinct orientations, there are 6^9 different orientations for each ordering. Each grid is found eight

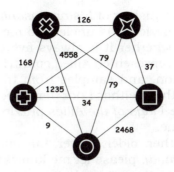

Figure 16. Graph for *On-the-spot Insanity*.

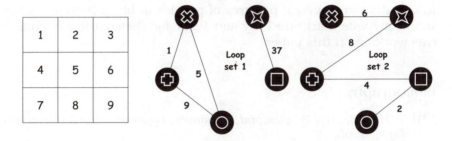

Figure 17. The solutions to *On-the-Spot Insanity* and its two loop sets

times and can be mirrored over the diagonal. Thus, the total number of distinct positions is $(6^9 \cdot 9!)/(8 \cdot 2)$. *On-the-Spot Insanity* thus has 5,511,240 times as many distinct positions! In comparison, *Osiris* has 11,612,160 distinct positions or 260 times as many distinct positions as *Instant Insanity*.

Open Issues

Clearly the graphical method is not the ideal method for solving *On-the-Spot Insanity*. I would be interested in methods invented by readers. It took about 50 years before the graph method was found for *Instant Insanity*. Hopefully it will not take that long for another method to arise. Note that, in general, *Instant Insanity*-type puzzles get harder with more cubes. Robertson and Munro [2] proved that *Generalized Instant Insanity*, with an arbitrary number of cubes, becomes computationally intractable as the number of cubes grows (formally, the puzzle is NP-complete).

Going through the history of *Instant Insanity*, I came across the *Flags of the Allies* puzzle. It is unclear to me how many versions of the puzzle exist, whether it is always five cubes (or sometimes four), which are the represented Allied countries (sources give contradictory information, and samples from the collection of Jerry Slocum show flags that neither I nor Jerry recognize). Reliable information about the *Flags of the Allies* puzzle and similar puzzles is more than welcome.

If you know of other, older, newer, but above all different versions of *Instant Insanity*, please let me know. I would like to learn their details: configuration, size, photo, origin, etc.

Acknowledgments. I thank Jerry Slocum for providing background information and several pictures of puzzles in his collection. I also thank my wife Maria and my son Peter for their patience while I was working on this paper.

Bibliography

[1] T. H. O'Beirne. *Puzzles and Paradoxes*. London: Oxford University Press, 1965.

[2] Edward Robertson and Ian Munro. "NP-completeness, Puzzles, and Games." *Utilitas Mathematica* 13 (1978), 99–116.

[3] David Singmaster. *Chronology of Recreational Mathematics*. Available at http://anduin.eldar.org/~problemi/singmast/recchron.html, 1996.

[4] David Singmaster. *Queries on "Sources in Recreational Mathematics."* Available at http://anduin.eldar.org/~problemi/singmast/queries.html, 1996.

[5] Jerry Slocum and Jack Botermans. *Puzzles Old and New*. Seattle: University of Washington Press, 1986.

[6] Rüdiger Thiele. "Katzenjammer Puzzle." In *Spiele, Spaß und Strategien—Die Gefesselte Zeit*. Velten: Reinhardt Becker Verlag, 1984.

Other Sources on Instant Insanity

The reader may be interested in these other sources:

[7] F. O. Armbruster. *Armbruster Puzzles Dot Com*. Available at http://www.armbrusterpuzzles.com/, 2003.

[8] Elwyn Berlekamp, John Conway, and Richard Guy. "The Great Tantalizer." In *Winning Ways for Your Mathematical Plays*, Second Edition, Volume 4, pp. 892–895. Wellesley, MA: A K Peters, 2004.

[9] Jürgen Köller. "MacMahon's Coloured Cubes." Available at http://www.mathematische-basteleien.de/macmahon.htm, 2002.

[10] Ivars Peterson. "Averting Instant Insanity." *MathTrek*. Available at http://www.maa.org/mathland/mathtrek_8_9_99.html, 1999.

[11] Jaap Scherphuis. "Instant Insanity/Buvos Golyok/Drive Ya Crazy." Available at http://www.geocities.com/jaapsch/puzzles/insanity.htm, 2007.

The Adventures of Ant Alice

Peter Winkler

Ants, even in a one-dimensional environment, are a source of fascination for amateur puzzlists and mathematicians. Presented here are ten puzzles (devised by myself, except where noted) involving our "favorite ant" Alice. Each puzzle is intended to illustrate some mathematical idea.

We begin with the basic "ant puzzle."

Falling Alice

Twenty-five ants are placed randomly on a meter-long rod, oriented east-west. The thirteenth ant from the west end of the rod is our friend, Ant Alice. Each ant is facing east or west with equal probability. They proceed to march forward (that is, in whatever direction they are facing) at 1 cm/sec; whenever two ants collide, they reverse directions. How long does it take before we can be certain that Alice is off the rod?

Guessing the End

When Alice does fall off, what is the probability that she falls off the end that she was originally facing?

Last One Off

What is the probability that Alice is the last ant to fall off the rod?

Counting Collisions

During the process, what is the expected (i.e., average) number of collisions that take place on the rod?

Damage to Alice

What is the expected number of collisions that Alice herself has?

Alice's Insurance Rate

What is the probability that Alice has more collisions than any other ant?

Damage to the Rest of the Ants

Suppose Alice has a cold, which is transmitted from ant to ant instantly upon collision. How many ants will be infected, on average, before the rod is cleared?

Alice at the Midpoint

Let us do a new experiment. Alice is carefully placed at the exact center of the meter rod, with 12 ants placed randomly to her west and 12 more to her east. As before, each ant faces randomly east or west, and they all march in whatever direction they are facing at 1 cm/sec, reversing directions whenever any two meet head-on. This time, however, ants do not fall off the rod: they turn around when they reach the end. One hundred seconds later, the ants are frozen in place. What is the maximum distance that Alice can be from her initial position?

Alice's New Whereabouts

There are only 24 ants on the rod now, with 12 on the west half facing east and the rest on the east half facing west. Alice is the fifth ant from the west end. The ants proceed to march as usual, reversing when they collide and falling off the ends. What do you need to know about the initial configuration in order to predict where Alice will be after 63 seconds?

Alice on the Circle

Now Alice is one of 24 ants randomly placed on a circular track of length 1 meter. Each ant faces randomly clockwise or counterclockwise and marches at 1 cm/sec; as usual, when two ants collide, they both reverse directions. What is the probability that, after 100 seconds, Alice finds herself exactly where she began?

Solutions

Falling Alice

The key to this (and succeeding) puzzles is to notice that if ants were interchangeable, it would make no difference to the process if they passed one another instead of bouncing. Then, it's clear that each ant is simply walking straight ahead and must fall off within 100 seconds. Because all the ants are off in 100 seconds, Alice in particular has fallen off as well.

A nice way to think about the puzzle, which avoids making the ants anonymous, is to imagine that each carries a flag. When two ants meet and bounce, they exchange flags. Thus, at all times each ant is carrying *some* flag, and the flags march straight past one another. When all the flags are off the rod, all the ants are off as well.

If you start an ant facing east at the west end of the rod, you can arrange it so that Alice ends up carrying her flag off the east end of the rod 100 seconds later. So waiting 100 seconds is both necessary and sufficient to clear the rod.

As far as I know, the first publication of the puzzle was in Francis Su's "Math Fun Facts" web column at Harvey Mudd College; Francis recalls hearing it in Europe from someone he can't trace, named Felix Vardy. The puzzle then showed up in the Spring/Fall 2003 issue of *Emissary*, the magazine of the Mathematical Sciences Research Institute (Berkeley, California).

Dan Amir, a former Rector of Tel Aviv University, read the puzzle in *Emissary* and posed it to T.A.U. mathematician Noga Alon, who brought it to the Institute for Advanced Study; I first heard it from Avi Wigderson of the I.A.S., in late 2003.

Guessing the End

The number of ants falling off the east end of the rod is the same as the number of ants facing east at the start, because the number of ants facing east never changes. (Alternatively, you can think of flags falling off the rod instead). In any case, if k ants fall off the east end, it is exactly the k easternmost ants who do so, because the ants stay in order.

We may assume, by symmetry, that Alice is facing east at the start, and we know she goes off the east end just when the number of east-facing ants is at least 13. This means that 12 or more of the *other* 24 ants are facing east. Of course, the probability that 13 or more of 24 ants face east is the same as the probability that 11 or fewer face east, so the probability of the event we are interested in is one-half, plus half the probability of exactly 12 of 24 ants facing east. The latter is $\binom{24}{12}/2^{24}$, which works out to 0.161180258; thus the answer is 0.580590129..., a bit over 58%.

Last One Off

We may assume (by symmetry, again) that Alice departs by the east end of the rod, which means that the 12 ants to her east do the same. If she is last off, it must be that the 12 ants west of Alice drop off the west end; it follows that initially exactly 12 flags, and thus exactly 12 ants, faced west. This happens with probability $\binom{25}{12}/2^{24}$, about 31%.

However, Alice is not necessarily the last ant off in these cases; about half the time, her western neighbor has the honor. Thus the desired probability is about 15.5%.

But are you satisfied with an approximation? Not when the exact answer is available. The time each flag is fated to spend on the rod is uniformly random, so the probability that the longest-lived flag is one of the 13 east-facing flags is 13/25. Hence the correct value is $13/25 \times \binom{25}{12}/2^{24}$, which is the same as the now-familiar number $\binom{24}{12}/2^{24}$, about 16.1180258%.

Counting Collisions

Each flag crosses all others ahead of it that are headed towards it; for the average flag, which starts at the midpoint of the rod, this is 6 of the 12 ahead of it. So the average flag hits 6 others; thus there are $25 \times 6 = 150$ "hits" on average. But this counts each collision twice, so the answer is 75.

An alternate, and slightly more rigorous, way to compute this is as follows. What is the probability that two flags cross? No matter where they are, this happens if and only if they face one another, so the probability is $1/4$. By linearity of expectation, then, the expected number of flag crossings is $\binom{25}{2} \times 1/4 = 25 \times 24/8 = 75$.

The maximum number of collisions is achieved if all ants are pointed toward Alice (the center ant), in which case all 13 flags facing Alice's way hit all 12 flags facing the other way, for a total of $12 \times 13 = 156$ hits.

The least possible number of collisions is of course zero, but this occurs with probability only $26/2^{25} \sim 0.000000774860382$.

Damage to Alice

It's easy to compute the number of collisions Alice's *flag* has: assuming (say) that Alice initially faces east, there will be an average of 6 (out of 12) flags ahead of Alice facing west, so her flag expects to pass six other flags.

But Alice is not always carrying her original flag, and in fact we expect Alice to have many more than 6 collisions on average. Why? Because the *average* ant has 6 collisions ($75 \times 2/25$) and Alice, being the middle ant, should have more than average.

Now, any given ant collides only with its two neighbors and alternates between them (because its direction alternates between collisions). An ant's *last* collision will be with its western neighbor if it ends up falling off the east end, and with its eastern neighbor if it goes off the west end.

Suppose that k ants face west initially. Because their flags march off the west end of the rod, the k westernmost ants end up dropping off the west end. Each of these ants who faced west initially will have an equal number of collisions on each side; those who faced east will have one extra collision on the east side. Thus, the number of collisions between ant j (counting from the west) and ant $j+1$ is equal to the number of east-facing ants among ants 1 through j—as long as $j < k$.

By symmetry, we may assume that k is between 13 and 25 (i.e., Alice herself drops off the west end). Then, the number of collisions between Alice and her western neighbor is exactly the number of east-facing ants west of Alice; call this number x. The total number of collisions experienced by Alice would then be $2x$ or $2x+1$, depending on whether she herself faced west or east at the start.

A priori, the expected value $E[x]$ of x is 6, because there are 12 ants west of Alice and each could face either way. However, we just assumed (darn!) that more than half the ants faced west. Note that, because the expected number of east-facing ants *east* of Alice is the same as $E[x]$, the number $2E[x+1]$ that we seek is exactly the total expected number of east-facing ants, given that they are in the minority.

Suppose that the ants were assigned directions in alphabetical order, with Ant Zelda last. There are $2^{25}/2 = 2^{24}$ ways to do the assignment so as to get a west-facing majority; of those, $\binom{24}{12}$ result in 12 east-facers among the first 24 choices. In those, Zelda is forced to face west; in the rest, she is equally likely to face west or east. It follows that the probability that she faces east is $1/2 - (1/2) \times \binom{24}{12}/2^{24} \sim 0.419409871$.

Because Zelda's probability of facing east is no different from any other ant's, we can multiply this by 25 to get the expected number of east-facing ants, about 10.4852468. This, then, is the average number of collisions experienced by Alice.

Alice's Insurance Rate

Suppose that the westernmost k ants fall off the west end, and the rest fall off the east end. We have seen in the previous puzzle solution that, if c_i is the number of collisions between ant i (counting from the west end) and ant $i+1$, then c_i stays the same or increases by 1 up to $i = k$; after that, c_i stays the same or decreases by 1. In particular, $c_i = c_{i-1}$ exactly when ant i faces (initially) the end from which he or she is fated to drop.

The number of collisions experienced by ant i is $c_{i-1} + c_i$, so in order for Alice to win the collision game, we need that $c_{11} + c_{12} < c_{12} + c_{13} > c_{13} + c_{14}$, which means that $c_{11} < c_{13}$ and $c_{12} > c_{14}$. This can only happen if $c_{11} < c_{12} = c_{13} > c_{14}$, which requires three properties: $k = 12$ or 13, Alice faces the end from which she drops, and her two neighbors face *away* from the ends from which they drop. This sounds like probability

$$\left(\binom{25}{12} + \binom{25}{12}\right)/2^{25} \cdot \left(\frac{1}{2}\right)^3 \sim 3.87452543\%,$$

but the events are not quite independent.

Suppose that Alice faces east, her eastern neighbor is Ed, and her western neighbor is Will. Then, Ed, like Alice, will be dropping off the east end and must then be facing west (probability $1/2$). Will will be one of the 12 ants dropping off the west end and hence must be facing east (probability $1/2$). The remaining 22 ants must be facing half west, half east (probability $\binom{22}{11}/2^{22}$), so the accurate answer is

$$\left(\frac{1}{2}\right)^2 \cdot \binom{22}{11}/2^{25} \sim 4.20470238\%.$$

Damage to the Rest of the Ants

Like many of the Ant Alice puzzles, this one is purely combinatorial. For example, somewhat counter-intuitively, it has nothing to do with the length of the rod. You might think that a shorter rod could enable some ants to get off the rod before they have a chance to become infected, but once an ant is headed for the end with no oncoming ants ahead, its collision days are over.

Probably the easiest way to make the required calculation is to think of flags being infected instead of ants. We may assume that Alice faces east; then all west-facing flags ahead of her will cross hers and become sick, while east-facing flags ahead of her will escape uninfected. In the meantime, the west-facing flags, after crossing Alice's flag, will infect all east-facing flags *behind* Alice, while the west-facing flags behind Alice get away.

Because there are an average of 6 west-facing flags ahead of Alice and 6 east-facing flags behind her, this seems to give an average of 13 infected flags (counting Alice's), and thus 13 infected ants.

However, there's a slight glitch: if there are *no* west-facing ants ahead of Alice, then there is no flag to cross Alice's and infect the east-facing flags behind her. This happens with probability $1/2^{12}$ and reduces the expected number of infectees from 7 (Alice plus an average of 6 east-facing flags behind her) to 1 (Alice alone). So the correct answer is not 13 but $13 - 6/2^{12} \sim 12.9985352$ sick ants on average.

Alice at the Midpoint

Suppose, as usual, that each ant carries a flag and that flags are exchanged when two ants meet. Then, each flag travels exactly one meter, bouncing once off the end of the rod and ending at a position symmetrically opposite its initial position. In particular, Alice's flag ends up back in the center. But will Alice be carrying it?

Indeed she will, because the ants remain in their original order. The 12 flags originally on the west side of the rod are now on the east side and vice versa, so Alice's flag is once again the 13th flag, and Alice herself is still the 13th ant.

So Alice ends exactly where she began; in other words, the maximum distance she can be from her starting spot is zero.

Contributed by John Guilford, of Agilent Inc., to Stan Wagon, who made it the Macalester College Problem of the Week.[1] I heard it from Elwyn Berlekamp, at the Joint Mathematics Meetings in Phoenix, January 2004. It was there that the central character in this paper received her name; I believe that Elwyn actually has an Aunt Alice. I was influenced as well by the presence at the conference of Alice Peters of A K Peters Ltd., publisher of my puzzle book, *Mathematical Puzzles: A Connoisseur's Collection* (2003).

The puzzle was reprinted in the Spring 2004 issue of *Emissary*.

Alice's New Whereabouts

Let $x_1, \ldots, x_{12}$ be the initial positions of the 12 west-facing ants, numbered from west to east; the positions are measured in centimeters from the west end of the rod. Let k be the smallest number such that the flags beginning at $x_{k+1}, \ldots, x_{12}$ remain on the rod, ending, therefore, at $x_{k+1} - 63, \ldots, x_{12} - 63$.

The ants, of course, remain in order. Because k flags drop off going west, Alice is gone from the rod if $k \geq 5$. Otherwise she is the $(5 - k)$th remaining ant, counting from the west end, which puts her in position $x_{k+(5-k)} - 63 = x_5 - 63$.

Thus, all you need to know is the position x_5 of the fifth ant on the east half of the rod, i.e., the 17th ant from the west end. Alice will end up 63 cm west of that spot; if that spot was already less than 63 cm from the west end, she falls off the rod.

This is a variation of a puzzle that was devised by Noga Alon and Oded Margalit of Tel Aviv University and communicated to me by Noga.

---

[1] http://mathforum.org/wagon/fall03/p996.html

As usual, we suppose that each ant carries a flag and that flags are exchanged when two ants meet. Then, each flag travels exactly once around the track in the given time period, ending where it began. The ants themselves must remain in the same circular order in which they started, so they have experienced (at most) some rotation: every ant must move the same number of positions, say k positions clockwise. In particular, Alice returns to her initial position only if all the ants do.

Note however that, if m ants are initially facing clockwise, then at any time there are always m ants moving clockwise and $24 - m$ moving counterclockwise. This is because, at each collision, a clockwise ant is exchanged for a counterclockwise ant; or, you can think of it as conservation of angular momentum! In any case, the average ant moves $m - (24 - m) = 2m - 24$ cm clockwise during the experiment. Thus, we are back to the initial position if and only if $2m - 24$ is a multiple of 24, i.e., if $m = 0$, 24, or 12.

The first two possibilities (where all ants initially face the same way) have negligible probability, but the last contributes a healthy 16.1180377%.

To be precise, there are $2^{24} = 16{,}777{,}216$ ways to choose directions for the ants, of which $\binom{24}{0} + \binom{24}{12} + \binom{24}{24} = 1 + 2{,}704{,}156 + 1$ bring Alice back to where she started. This gives probability $2{,}704{,}158/16{,}777{,}216$.

Part IV

Fitting In

Simplicity

Stewart Coffin

One hundred and sixty years ago, the iconoclastic hermit of Walden Pond offered this advice: "simplify, simplify!" One can only wonder what Thoreau would have thought today. My new cell phone came with 50 pages of instructions, and Lasser's latest income tax guide has over 800 pages, most of which are incomprehensible to me. Do things really need to be that complicated? Our natural yearning for simplicity might also apply to mathematical recreations, and more specifically to my special interest, which is geometrical and mechanical puzzles. Historically, it has been the simpler amusements that have usually enjoyed the most enduring popularity. Tinkertoys and building blocks are likely to still be around long after all of today's video games have been discarded.

Of course, to the aspiring inventor, it always seems as though most of the simpler ideas have long ago been conceived and brought to light, perhaps even copyrighted or patented. All the more satisfaction, then, when the explorer of ideas stumbles upon a simple amusement that appears to be new and original, as much as anything in this world can truly be so described. For this article, I have sifted though my 35-year accumulation of puzzle designs and picked out a few that best illustrate the concept of simplicity.

Some of my more satisfying geometrical dissection puzzles have involved fitting four pieces into a square or rectangular tray, while

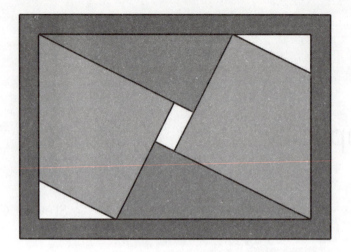

Figure 1. A peudo-dissection puzzle.

leaving some empty spaces. Because traditional dissection puzzles involve pieces packing solidly to fill all gaps, these puzzles are perhaps more properly called *pseudo-dissection* puzzles. Shown in Figure 1 is one that Mary and I frequently take on Elderhostel trips to entertain our companions, very few of whom are ever able to solve it, even after receiving hints. This puzzle exploits one's overpowering natural tendency to begin by trying to fit square corners of pieces into square corners of the container, as we all have been doing habitually, practically from birth. Even when the hapless victims of this psychological trap are cautioned to try a different approach, they will almost invariably revert to this hopeless first step.

My most successful polyomino-type puzzles have involved fitting just four or five pieces into a square or rectangular tray, as in the example shown in Figure 2. The size of the square tray is such that the five pieces fit snugly when they are arranged symmetrically as shown. Again, the success of the design is based more on psychology than mathematics. We spend most of our lives immersed in a world of orthogonal arrangements—everything from city streets and building plans to printed pages and computer screens. Thus the unfortunate puzzle solver has much difficulty ignoring this predisposition long enough to place the first piece skewed at an angle to the tray. I call this general class of puzzles "Square Root Type," and I call this subclass "Square Root of Five," referring to the rel-

Figure 2. A "Square Root of Five" puzzle.

Figure 3. A "Square Root of Ten" puzzle.

ative dimensions of puzzle pieces and tray. Closely related is the "Square Root of Ten" subclass, perhaps even more confusing, as in the example shown in Figure 3. Many other variations on this theme are possible.

In the world of cubic puzzles, the $3 \times 3 \times 3$ size was popularized by Piet Hein's seven-piece *Soma Cube*. The puzzle is based on dividing a cube into 27 equal parts according to a $3 \times 3 \times 3$ grid and joining these parts to form the puzzle pieces. Perhaps because of the multiple solutions (over 200) for Piet Hein's puzzle, the 27-block size is often overlooked by puzzle designers as more of a novel plaything rather than a real challenge. Indeed, the tendency among puzzle designers these days (myself included) has been to tinker with interlocking assemblies of greater size and complexity. But of my designs in this class, my favorite is still the classic $3 \times 3 \times 3$ *Half-Hour Puzzle* (see Figure 4). Here, my design objective was to discover a set of pieces, all dissimilar and asymmetrical, preferably all the same size, and with the maximum number of pieces that would assemble into a $3 \times 3 \times 3$ cube in only one way. Because some of these requirements are mutually exclusive, a compromise is required, resulting in the six pieces shown below. As the name suggests, a half hour is a reasonable time for discovering the one solution. Incidentally, many puzzle solvers now have access to computer programs that can solve puzzles of this sort with blinding speed, by a process that might be described as systematic trial and error. One of the benefits of solving such puzzles the old-fashioned

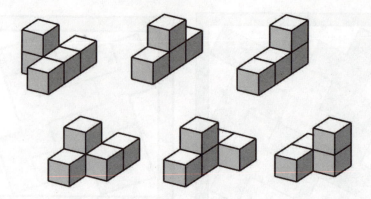

Figure 4. *Half-Hour Puzzle.*

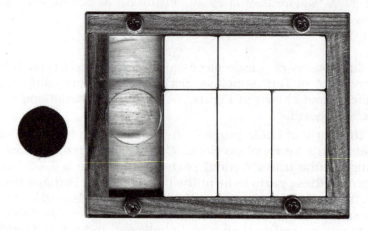

Figure 5. *Drop Out Puzzle.*

way is that nothing the human brain does can truly be described as random. One is continually discovering educated tricks and clever shortcuts to the solution, whether consciously or otherwise, and that can be a recreation in itself.

In 1964, Martin Gardner wrote a *Scientific American* column entitled "The Hypnotic Fascination of Sliding Block Puzzles".[1] In it he discussed the classic nine-piece *Dad's Puzzle*, a particular favorite of my childhood, although I don't believe I ever solved it. In these

[1]*Scientific American*, 210:122–130, 1964. Reprinted in Martin Gardner's *Sixth Book of Mathematical Diversions from Scientific American*, Chapter 7, pages 64–70 (University of Chicago Press, 1983).

many years, the hypnotic fascination hasn't changed. Some of the recently published designs of sliding-block puzzles are exceedingly clever and complex, especially the three-dimensional ones. Even among the flat kind, most are larger and more complex than the 4×5 *Dad's Puzzle*. But I wondered: are simpler designs possible? This question led to the creation of the *Drop Out Puzzle*, shown in Figure 5. There are six movable pieces. The 3×4 tray has a clear plexiglass top with a circular hole at one end, through which the round disk can be dropped. The object is, by shifting the pieces about, to eventually drop the disk through a hole in the bottom of the tray at the opposite end. (An additional hole in the center of the cover, not shown, is merely to facilitate pushing the pieces around with the eraser-end of a pencil.) It may appear to be impossible, but it can be done. I don't think you will find a puzzle much simpler than this. I wonder if Thoreau would have approved.

Extreme Puzzles

Frans de Vreugd

For many years I have been interested in mechanical puzzles. One of the types of puzzles I like best is interlocking puzzles. For these puzzles, several pieces have to be moved before you can remove the first piece from the puzzle. In the last 20–30 years, there have been many developments in this category of puzzles. It seemed like a worldwide, ongoing "race" to find the puzzle needing the highest number of moves to remove the first piece. These puzzles are often referred to as *high-level* puzzles. In the past ten years, many puzzle designs have appeared, requiring stunning numbers of moves to take out the first piece. This article describes many of these *extreme puzzles*.

History

One of the classic interlocking puzzles is the *Chinese Cross* (a.k.a. *Six-Piece Burr*); see Figure 1. Bill Cutler made a complete study of this type of puzzle using dedicated software. The standard version of this puzzle is relatively easy to solve. A solid piece is removed first, and then the rest would follow quickly. Most of these puzzles are solid. Once assembled, there are no voids inside the puzzle. Bill Cutler searched for puzzles that would take more than one move to take the first piece out. Having empty spaces inside the

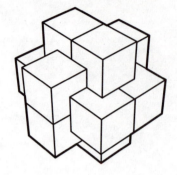

Figure 1. *Chinese Cross.*

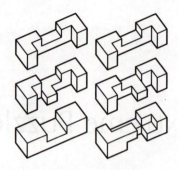

Figure 2. Pieces of *Bill's Baffling Burr.*

puzzle is a requirement to have more than one move for removing the first piece (or pieces). The internal voids allow pieces to be moved without being taken out of the puzzle.

An amazing puzzle that Bill Cutler found with the aid of his program was the first high-level puzzle published (in *Scientific American*), known as *Bill's Baffling Burr* (see Figure 2). At the time it was absolutely amazing that it took no less than five moves to remove the first piece. Several years later Bruce Love designed a similar six-piece puzzle that needs 12 moves to remove the first piece, appropriately named *Love's Dozen*; see the pieces of this puzzle in Figure 3. The computer research done by Bill Cutler has proved that this is in fact the highest level possible with six-piece burrs.

In 1958 Willem van der Poel from Zoetermeer, the Netherlands, introduced another classic interlocking puzzle: the *Van der Poel Puzzle* (Figure 4). It consists of 18 pieces, subdivided into a cage of 12 identical pieces and an internal lock of six other pieces. The puzzle is not extremely difficult (it needs only a few moves to get the first piece out), but the introduction of the outside cage with the internal lock proved to be an inspiration for many other high-level puzzle designs.

It was again Bruce Love who came up with a stunning record. He designed an 18-piece puzzle that needed no less than 18 consecutive moves to get the first piece out. The pieces of this puzzle are shown in Figure 5. For many years the record of 18 moves was unchallenged. Then, in the nineties, several puzzlers from all over the world came up with improvements, which will be discussed in this article. I will start discussing puzzles with many pieces and continue with fewer and fewer pieces.

A Lifetime of Puzzles

Figure 3. Pieces of *Love's Dozen*.

Figure 4. *Van der Poel Puzzle*.

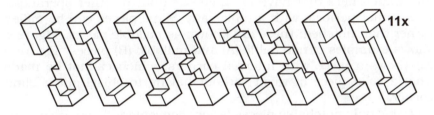

Figure 5. Pieces of *Lovely*.

Design Criteria

The number of moves you need to get the first piece out of puzzle (called the *level* of a puzzle) is a good indicator of the level of difficulty of the puzzle. A *level 7-4 puzzle* means that it requires seven moves to get the first piece out and another four moves for the second piece. There are more criteria that make a puzzle interesting. It is nice to have a puzzle requiring many moves to solve, but it is even nicer if there is only one way the puzzle goes together. If you have a set of pieces that makes a level-18 solution but has many alternative (and very easy) solutions, this is usually not a good thing. If the pieces go together in only one way, it is called a *unique solution*. Apart from looking for high-level puzzles, designers also tend to look for these unique solutions.

When you look at the pieces of a high-level puzzle, they tend to be rather complex. Producing a puzzle like that might be either difficult, expensive, or both. Therefore, puzzle designers often look for pieces with a special characteristic, called *notchable pieces*. A notchable piece can be made with just a table saw. There are no

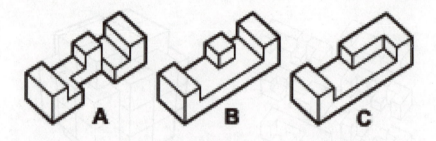

Figure 6. (A) Notchable, (B) millable, (C) non-notchable pieces.

internal corners that have to be chiseled out or other operations you cannot perform with a table saw. Some pieces can be milled rather than notched; these are known as *millable pieces*. Figure 6 shows examples of notchable (A) and millable (B) pieces. Figure 6 (C) shows a piece with internal corners, which cannot be made from a solid piece of wood without having to chisel out a blind corner.

Using only notchable pieces is very convenient if you are manufacturing a puzzle. Another design criterion might be to have many identical pieces.

Eighteen-Piece Puzzles

For many years it was thought that high-level puzzles were possible only if you had a puzzle with many pieces. The 18-piece puzzle proved an important source of inspiration for puzzle inventors worldwide. In the late 1990s, Brian Young from Tamborine, Australia, reported his own design, needing 19 moves to get the first piece out! This puzzle, called *Coming of Age Mark II*, is still a best seller in his puzzle business, Mr. Puzzle. Brian did not use a computer for this design.

The help of the computer meant an enormous leap forward in creating high-level puzzles. Computer programmer Pit Khiam Goh from Singapore designed a puzzle called *Burrloon*. The record of 19 moves from Brian Young was beaten, and not by just one or two moves. To remove the first piece from the *Burrloon* puzzle requires no less than a staggering number of 33 moves. Moreover, this puzzle has two very interesting characteristics: the solution is

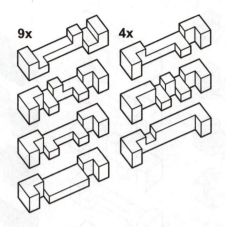

9x   **4x**

Figure 7. Pieces of *Burrloon*.

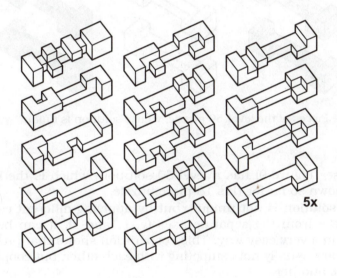

5x

Figure 8. Pieces of *Tipperary*.

unique, and all the pieces are notchable. The pieces of *Burrloon* are shown in Figure 7.

In 2003, Jack Krijnen, a puzzle designer from the Netherlands, improved the record again. His puzzle *Tipperary* has a wonderful unique level-43 solution. The pieces are shown in Figure 8.

Two years later, Jack Krijnen and Pit Khiam Goh joined forces and came up with another record-breaking design. This puzzle (for

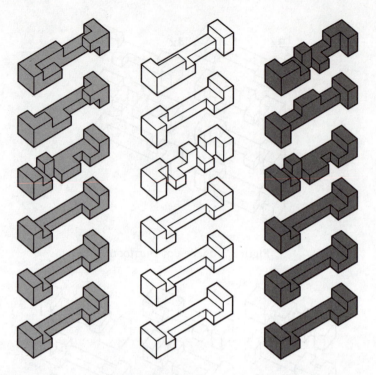

Figure 9. Pieces of the level-50 puzzle, whose solution is unique with three colors.

pieces see Figure 9) has a level-50 solution, which is the highest level known so far for this 18-piece puzzle.

The solution is not unique, but by applying different colors to the pieces from the separate axes (x, y, and z), it can be made unique in a very easy way. This cooperation shows that puzzle designers are usually not competing with each other, but cooperating with one another.

Fewer Pieces

Dic Sonneveld from the Netherlands has been a pioneer in the field of high-level puzzles. From the early days on, he has been fine-tuning many designs in order to try to raise the level of many of his own designs. He uses an interesting technique for this. He starts with an existing solution for a puzzle and analyzes where the open spaces in the puzzle are. Then, he adds a cube to one of these positions and tries to disassemble the puzzle again. In several cases this resulted in a higher level. Although this method worked, it was an arduous task to analyze all the different solutions, empty positions, and the new resulting puzzles. Often it took many weeks to get just one extra move in the solution! Dic noticed that the approach seemed to work, but that he really needed a computer to do all the hard work. He developed a computer program that started as a Word macro but soon turned into a complex Visual Basic program. This program finds all the empty positions for every single solution of the puzzle and generates new puzzles from that. Then, all new puzzles are recalculated (to find out how many moves it would take to disassemble these new puzzles), and then the previous steps are repeated. As a test project, he used one of his designs called *Dic's Dozen*, a twelve-piece puzzle with an interesting symmetry pattern; see Figure 10.

The results from his computer program were truly breathtaking. By hand he had never come any further than 12 or 13 moves, which seemed quite good at the time. When the program got up and running, however, it started to pop out solutions with incredible numbers of moves to remove the first piece. His first result needed no less than 31 moves. That seems a lot less than the 43 of the *Tipperary* design, but *Dic's Dozen* used only 12 pieces instead of the 18 of *Tipperary*. And this was only the start. The computer ran for hours and hours, resulting in higher levels time after time.

Soon after the 31-move version, a 39-move version was found. A week later the record was beaten again: 49 moves! And this was not the end yet. A few weeks later, another leap forward was reported: the computer had found a puzzle of level 64. (See Fig-

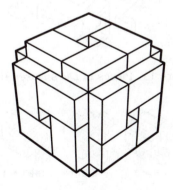

Figure 10. *Dic's Dozen.*

ure 11.) It is quite difficult to imagine having a structure of 12
pieces of wood for which you need 64 consecutive moves to get a
single piece out. And these were not the easiest moves. You had to
move five pieces in one direction, then three in another, then four
in another direction, etc.; see Figure 12 for an example of the complicated moves. Actually, the research got a little bit out of hand.
It started as a theoretical exercise to find an interesting puzzle, but
in the meantime it had resulted in puzzles so horribly difficult that
no living human was ever able to solve these puzzles! Neverthe-
less, many people were interested in these record-breaking puzzles
and have had them made. I wonder how many actually found the
solutions themselves....

Six-Piece Puzzles

When it turned out that you could get high-level puzzles with 18-
and 12-piece puzzles, the question arose whether you could also
get high levels with fewer pieces. This usually adds to the attraction of a puzzle. If you have a puzzle with many pieces, people have

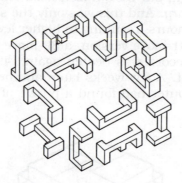

Figure 11. Pieces of *Dic's Dozen 64-5*.

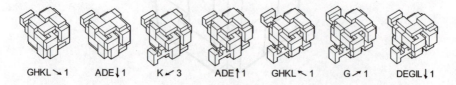

GHKL ↘ 1   ADE ↓ 1   K ↙ 3   ADE ↑ 1   GHKL ↘ 1   G ↗ 1   DEGIL ↓ 1

Figure 12. Complicated moves in the solution of *Dic's Dozen 64-5*.

A Lifetime of Puzzles

no trouble understanding that it might be very difficult to put together or take apart. However, if a puzzle has only six pieces, how hard could it be? Many different six-piece puzzles were designed in the past years that look very much the same on the outside, but have pieces that are very different. I will start with puzzles with complicated pieces and then move on to puzzles with simpler pieces.

Six Complicated Pieces

Lars Cristensen, a puzzle collector and designer from Denmark, has designed several high-level puzzles. One of his creations is called *Belle L-Burr* (see Figure 13), based on Kint-Bruynseels' *New L-Burr*. This puzzle has rather complicated pieces, as can be seen in Figure 14. Making a physical model of this puzzle is therefore not the easiest task. Once you have managed to make the pieces, though, you will have a very difficult puzzle. The pieces will go together in only one way, and once assembled, you need 40 moves to remove the first piece. Needless to say, it is a very, very difficult puzzle.

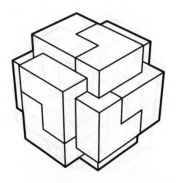

Figure 13. *Belle L-Burr.*

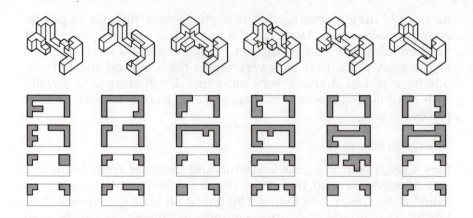

Figure 14. Pieces of *Belle L-Burr*.

Six Board Burrs

An interesting group of puzzles that was researched by Bill Cutler and myself is the so-called *Six Board Burrs*. These puzzles consist of six flat pieces of dimensions $1 \times 4 \times 6$, as shown in Figure 15. Junichi Yananose from Japan has done a lot of fieldwork and has designed many of these puzzles. *Six Board Burrs* have an extra attraction. Removing the first piece might not be as difficult as in other puzzles, but since the pieces are extremely interlocking, it often takes many moves to get the second, third, and further pieces out. In the research that Bill Cutler and I did, we considered not only the *regular* pieces (pieces derived from the basic C-shape), but also the *irregular* pieces; see Figure 16. We did a full analysis of all

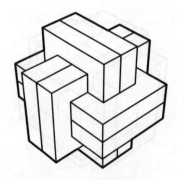

Figure 15. *Six Board Burr*.

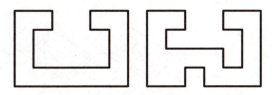

Figure 16. Examples of a regular piece (left) and an irregular piece (right).

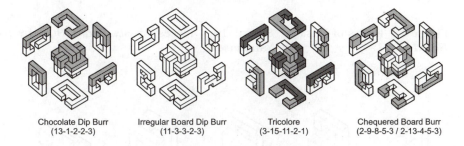

| Chocolate Dip Burr (13-1-2-2-3) | Irregular Board Dip Burr (11-3-3-2-3) | Tricolore (3-15-11-2-1) | Chequered Board Burr (2-9-8-5-3 / 2-13-4-5-3) |

Figure 17. Four different *Six Board Burr* designs.

combinations of all 219 possible puzzle pieces. In total, we found no less than 14,563,061,989 ways to fit six pieces together.

Some interesting designs are shown in Figure 17:

- *Chocolate Dip Burr* (highest level: level 13),

- *Irregular Board Burr* (unique level-11 solution),

- *Tricolore* (large number of moves for second and third pieces: 3-15-11-2-1),

- *Chequered Board Burr* (two solutions, both very hard: 2-9-8-5-3 and 2-13-4-5-3).

Before we started our analysis, some interesting designs already existed. Lars Christensen designed a six-board burr puzzle known as *Basic Board Burr* (Figure 18), which is regarded as one of the best in this category. By changing the length of the pieces from six units to eight units, you could make the puzzle even harder, as shown by the six-board burr in Figure 19 that was introduced by Junichi Yananose from Japan.

The nice thing about the six-board burrs is that the puzzle stays coherent even if you take a piece out. Sometimes a very specific sequence of moves is needed just to interchange the place of two

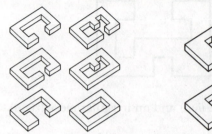

Figure 18. Pieces of *Basic Board Burr*.

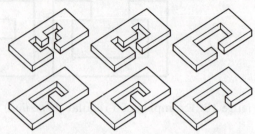

Figure 19. Pieces of *Six Board Burr* by Yananose.

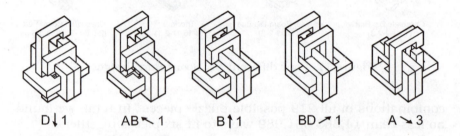

D↓1   AB↖1   B↑1   BD↗1   A↘3

Figure 20. Two pieces of *Tricolore* changing places by moving through each other.

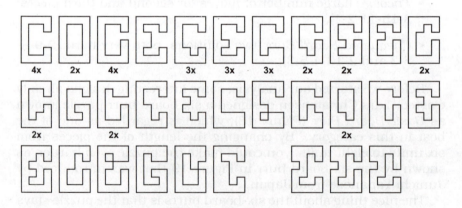

Figure 21. Pieces that make all high-level solutions for *Six Board Burr*.

pieces. An example of this is shown in Figure 20. This is an important mechanism to achieve high-level solutions.

From the research that Bill Cutler and I did, we started filtering out the nice solutions. It turned out that if you use a subset of only 31 pieces (out of the 219 possible pieces), you can make all puzzles having a first level of 8 or higher, a second level of 13 or higher, and/or a total number of moves for the first and second level of at least 16 moves. Figure 21 shows this set of pieces.

Separated Board Burrs

An interesting sidestep from six-board burrs was introduced by Jim Gooch from the United States. He suggested that it would be a nice idea to have the pieces not side by side, but separated by one unit; see Figure 22. These so-called *Separated Board Burrs* turned out to be very difficult. I used this puzzle to test the program Dic Sonneveld had written just for *Dic's Dozen*.

Shortly after he had done the research on this puzzle, Sonneveld rewrote the program for more general use. One of the first and fascinating results was *Zigzag*, a six-piece puzzle with three different pieces, two of each; see Figure 23. The distinct zigzag shape of two of the pieces explains the name of the puzzle. George Miller from Sonoma, CA, made some nice copies of this puzzle from laser-cut acrylic; see Figure 24. *Zigzag* was just the beginning: the potential for high-level puzzles in this group is enormous. It is interesting to see that a puzzle with only six flat pieces, which do not look too complicated, can result in a very difficult puzzle.

One of the most interesting puzzles of this type has a unique level-17 solution. Only six pieces, and still you need 17 moves to

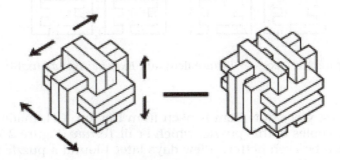

Figure 22. Transformation from *Six Board Burr* to *Separated Board Burr*.

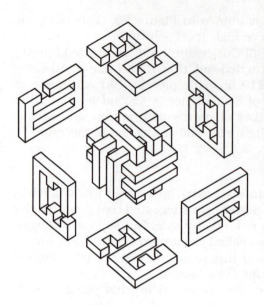

Figure 23. *Zigzag*.

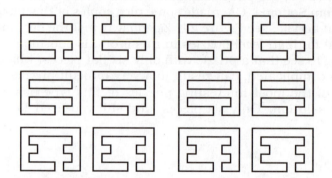

Figure 24. *Zigzag* made by George Miller.

Figure 25. Pieces of *Torture* (left) and *Extreme Torture* (right).

get the first piece out. Tom Lensch from Dayton, OH, made some very nice copies of this puzzle, which I call *Torture* (Figure 25, left). But, it can be even better: a few days later I found a puzzle much harder than the previous one. It needs no less than 28 moves to remove the first piece! I called this improved version *Extreme Torture* (Figure 25, right).

The total sequence for complete disassembly is 28-21-9-8-3. So, when you have removed the first piece after 28 moves, you need another twenty-one to get the second piece out. And even at the moment that only three pieces are left, you still need eight moves to get a piece out!

Bent Board Burrs

Another interesting variation on the six-board burrs is the so-called *Bent Board Burrs*. If you take the piece of a six-board burr and "bend" over one of the ends, you get L-shaped pieces; see Figure 26.

Initially, I thought that adding an extra "plate" to the end of the piece would severely restrict the moves, resulting in lower levels. This was not the case. For *Six Board Burrs*, the highest level found was a level-11 puzzle; for the *Bent Board Burrs*, this number literally doubled. The hardest *Bent Board Burr* has a level-22 solution. Figure 27 shows the puzzle (middle) and the pieces of this puzzle. Just like with the *Six Board Burrs* and the *Separated Board Burrs*, getting the second piece out can be just as hard as,

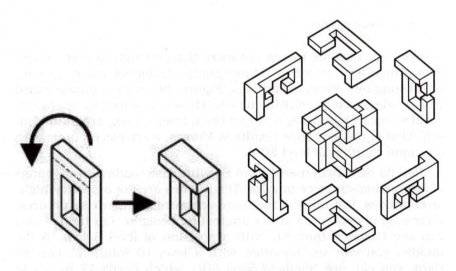

Figure 26. "Bent" pieces.   Figure 27. *Bent Board Burr #1.*

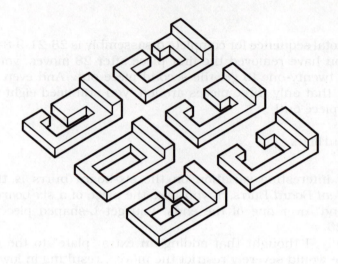

Figure 28. Pieces of *Bent Board Burr #4*, with solution of level 10-10.

or harder than, getting the first one out. A nice example is *Bent Board Burr #4*, where it takes ten moves to get the first piece out and another ten to get the second piece out. Figure 28 shows the pieces of this puzzle.

Other Six-Piece Puzzles

It turns out that six pieces are more than enough to make interesting high-level puzzles. Many puzzle designers came up with interesting new ideas for puzzles. Figure 29 shows a puzzle called *Mitosis*, designed by Pit Khiam Goh. He was inspired by Dic Sonneveld's research and decided to write a level-raising program himself. One of the very nice results is *Mitosis*, a six-piece puzzle with a unique solution of level 20-8.

Ronald Kint-Bruynseels from Belgium also made a whole range of interesting six-piece puzzles. His puzzles are not only very-high-level puzzles, but most of them are very nice geometric structures. Figure 30 shows three of his interesting designs. On the left you can see *Escher's Burr #1*, with a solution of level 3-13. In the middle, you can see *Squarrel*, with a level-10 solution. On the right, you can see *Sheffield Steel 6BB*, which needs 17 moves to take out the first piece and 14 more for the second.

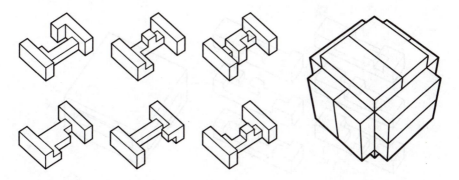

Figure 29. *Mitosis* and its pieces.

Figure 30. Three designs from Ronald Kint-Bruynseels: *Escher's Burr #1* (left), *Squarrel* (middle), and *Sheffield Steel 6BB* (right).

Boxed Burrs

An interesting group of interlocking puzzles is the so-called *boxed burrs* (or *framed burrs*). Apart from having the pieces themselves, there is a box in which they have to be fitted. Usually this means severe restrictions for the movement of the pieces. A nice example

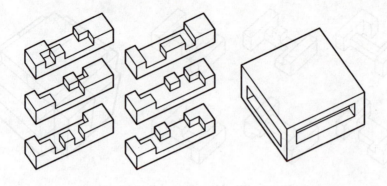

Figure 31. Pieces of *Framed Burr* by Yananose.

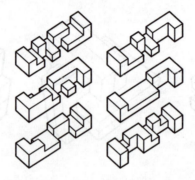

Figure 32. Pieces of *Fundamental Framed Burr* by Christensen.

is a six-piece framed burr by Junichi Yananose. He came up with the design shown in Figure 31. It has six pieces and a frame to fit them in. When the puzzle is assembled, it takes 17 moves to get the first piece out, and the solution is unique. Lars Christensen's *Fundamental Framed Burr* (Figure 32) has a level-27 solution; unfortunately, the solution is not unique.

Inspired by *Dic's Dozen* and Yananose's *Framed Burr*, I came up with a design for a boxed burr combining the qualities of both puzzles. The puzzle has only four pieces and a box, yet the results were amazing. After doing a few test runs with Dic's program, I noticed the enormous potential of this puzzle shape. So far, we had a record of 40 moves for a six-piece puzzle, meaning an average of 6.6 moves per piece. Could this new criterion be beaten? The answer was yes, although the number of pieces might be a bit confusing. Do you have four pieces and a box, or do you consider

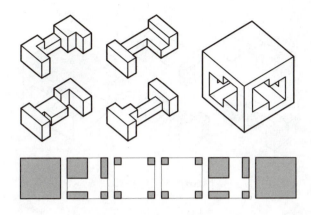

Figure 33. *Extreme Boxed Burr.*

the whole puzzle to be five pieces? Anyway, the record was broken, whichever way you count! One of the nicest results is a puzzle I call *Extreme Boxed Burr*, which needs 23 moves to free the first piece. It uses notchable pieces only (the box could be made from two identical notchable halves) and has a unique solution. The puzzle is shown in Figure 33.

A very similar design is called *Life@21* (see Figure 34). The pieces are very similar to the ones of *Extreme Boxed Burr*, but the solution is not unique. It has another nice characteristic, though. If you use two colors for the puzzle (which looks like a chocolate dip), the colors of the box and pieces will match only if you find the level-21 solution. In the other two solutions (being of levels 7 and 9 respectively), the colors do not match up. This puzzle was made commercially by Bits & Pieces in the US. It was presented assembled but with the colors not matching. It was the challenge for the puzzler to first get the pieces out and then to put them back in such a way that the colors would match.

There is another interesting variation. If you twist one end of a piece, you get what I called "twisted" pieces. This makes solving the puzzle quite confusing: on one end of the puzzle, you push a horizontally oriented piece in, and on the other side a vertically oriented piece comes out! Using these twisted pieces, some very high levels were found. The highest was a level-47 puzzle. Even if you count the box as a piece, it still means that you have an average of more than nine moves per piece! The puzzle had a problem, though. This is a problem you often run into with high-level puzzles: if you allow rotational moves, you can easily find a shortcut

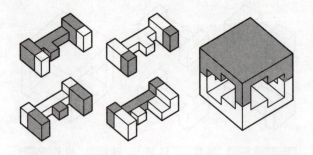

Figure 34. *Life@21.*

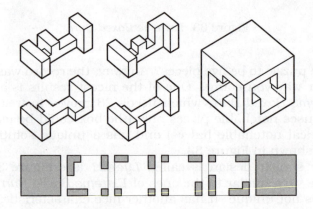

Figure 35. *Twisted Boxed Burr.*

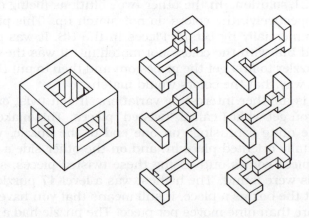

Figure 36. *The Pelican.*

to disassemble the puzzle with many fewer moves. In this case, the level 47 turned into a rather disappointing level 9 or so. For some other puzzles in this group, the rotational moves are less of a problem. An interesting design is the *Twisted Boxed Burr*, using four rather complicated pieces (see Figure 35). The move sequence is 29-22-6-3, resulting in 60 moves in total to disassemble the puzzle. Truly amazing is that when only two pieces are left in the box, you still need six moves to get one out.

Dic Sonneveld also designed an interesting boxed burr. The pieces are utterly complicated (not only to manufacture, but also to handle). In Figure 36 you can see the pieces of this puzzle, known as *The Pelican*.

Four Pieces

For an interlocking puzzle, four pieces seems an awkward number: very few four-piece interlocking puzzles are known. There is one nice example, though: a puzzle designed by Vesa Timonen, a talented Finnish puzzle designer. His *Vesa Burr 4* has four complicated pieces that intersect in only two directions; see Figure 37. It still needs 14 moves to remove the first piece, however.

Three Pieces Only?

As announced earlier, I am reducing the number of pieces as the article develops. *Grand Giga Burr*, designed by Lars Cristensen, is

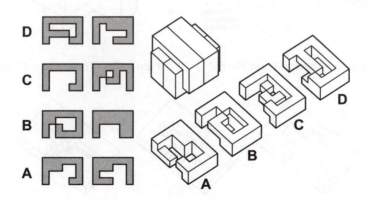

Figure 37. Vesa Timonen's *Vesa Burr 4*.

a nice example of a three-piece puzzle with a high-level solution. (See Figure 38.) The pieces are complicated, just like in *Belle L-Burr*, and the number of moves is remarkable. You need 19 moves to get the first piece out. If you allow rotational moves, you "only" need 16 moves.

Three Simple Pieces

It was again Jim Gooch who had an interesting idea to make a high-level puzzle with only three pieces. It is a variation on the well-known *Three-Piece Cross*; see Figure 39. It requires only three moves to get the first piece out and is therefore considered an easy puzzle. Jim Gooch came up with an idea to replace the inner 1×1 units by a 2×2 grid, allowing more complicated pieces. Figure 40 shows the pieces of one of my designs that needs eight moves to get the first piece out.

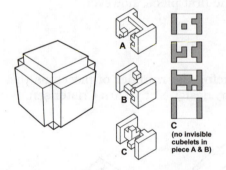

Figure 38. *Grand Giga Burr.*   Figure 39. *Three-Piece Cross.*

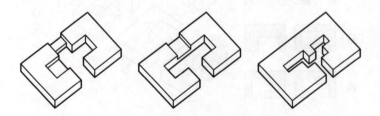

Figure 40. Pieces of *Three Piece NOT.*

Making Solutions Unique

If the solution of a puzzle is not unique (i.e., if there are more ways than one to assemble a puzzle), there are many ways to make it unique. What it basically comes down to is that you want to fix certain pieces into certain places or orientations. You could do this by using colors, as in *Life@21*. Sometimes it is enough to keep just two specific pieces together. In that case, making an imprint on these two puzzle pieces is enough. Other, more complex ways to make a solution unique are to make a slightly deformed version of your puzzle. You could squash it in one direction and stretch it in another; see Figure 41. In that case you have isolated the three main directions (you cannot use a y-direction piece in the x- or z-direction).

If this is not enough, you might consider making a slanted version of the puzzle, as shown in Figure 42. You can make it either slanted in one direction (left) or slanted in two directions (right). Stewart Coffin has made several slanted versions of well-known puzzles. Do not forget that this makes production of a puzzle much harder, because you are dealing with complicated compound angles. If the orientation has to be fixed completely, you could use dice as building blocks; see Figure 43. Another, rather sophisticated way is to interconnect the pieces, as can be seen in Figure 44.

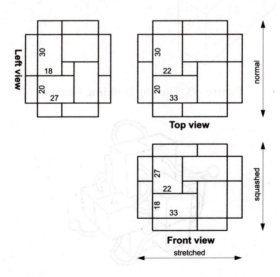

Figure 41. Stretched version of *Dic's Dozen*.

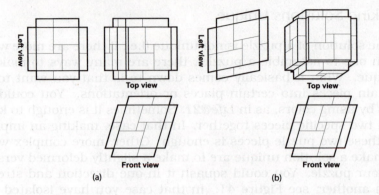

Left view

Top view

Left view

Top view

Front view

Front view

(a)

(b)

Figure 42. Slanted in (a) one direction and (b) two directions.

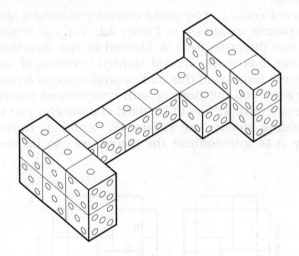

Figure 43. Dice as building blocks.

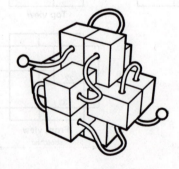

Figure 44. Another way to make a solution unique.

A Lifetime of Puzzles

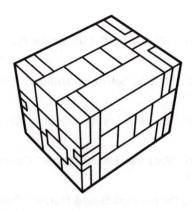

Figure 45. *Binary Burr*.

Conclusion

The ongoing "race" to find high-level puzzles has resulted in many interesting designs. In the early days, these designs were found by hand, and puzzle designers were mainly aiming at finding a higher level, independent of the number of pieces. When the computer got involved, there were many leaps forward in the developments. It was no longer just the number of moves that counted: getting a high number of moves with fewer pieces started to become the new goal. Other criteria, like finding unique solutions or having only notchable pieces, started to become more important.

Another interlocking puzzle worth mentioning is Bill Cutler's design for a *Binary Burr*; see Figure 13. This puzzle uses a mechanism equivalent to a classic puzzle known as the *Chinese Rings*. This puzzle is scalable. Every time you add one unit to the puzzle, you almost double the number of moves. Bill Cutler was able to translate the mechanism of this wire puzzle into an interlocking puzzle. His *Binary Burr* has 21 pieces and needs 85 moves before the first piece comes out. Since the system is scalable, the number of moves you might get is virtually unlimited!

Bibliography

[1] Edward Hordern. "What's Up?" *Cubism for Fun* 42 (February 1997), 31.

[2] Frans de Vreugd. "Making Solutions Unique." *Cubism for Fun* 53 (October 2000), 12–17.

[3] Frans de Vreugd. "Raising Levels." *Cubism for Fun* 55 (June 2001, 4–7.

[4] Frans de Vreugd. "Extreme Boxed Burrs." *Cubism for Fun* 56 (October 2001), 32–37.

[5] Frans de Vreugd. "Separated Board Burrs." *Cubism for Fun* 58 (July 2002), 35–39.

[6] Frans de Vreugd. "Cracking the Six Board Burr." *Cubism for Fun* 62 (Nov 2003), 18–23.

[7] Frans de Vreugd. "Analysis of the Three-Piece Cross II." *Cubism for Fun* 63 (2003), 31.

[8] Keiichiro Ishino. "Puzzle Will Be Played...." Available at http://www. asahi-net.or.jp/~rh5k-isn/Puzzle/index.html.en, 2000.

[9] John Rausch. "Puzzle World." Available at http://www.johnrausch. com/PuzzleWorld/, 2007.

Satterfield's Tomb

David A. Klarner
Wade Satterfield
edited by Thane E. Plambeck

Imagine a stack of 20 cannonballs in the form of a regular tetrahedron. The layers of such a stack are shown in Figure 1.

The number of balls in each layer are the so-called *triangular numbers*, $1, 3, 6, 10, \ldots$, having the form

$$1 + 2 + 3 + \cdots + L = \frac{L(L+1)}{2} = \binom{L+1}{2}.$$

The number of balls in a tetrahedral stack of L layers is the sum of the first L triangular numbers, giving rise to the *tetrahedral numbers*, $1, 4, 10, 20, \ldots$ for $L = 1, 2, 3, 4, \ldots$, respectively. In general,

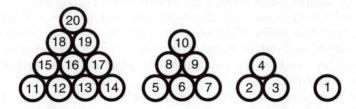

Figure 1. Twenty cannonballs in four layers.

Figure 2. The rhombic dodecahedron (interior cell).

the tetrahedral number corresponding to a stack with L layers will have

$$\binom{2}{2} + \binom{3}{2} + \cdots + \binom{L+1}{2} = \frac{(L+2)(L+1)L}{6} = \binom{L+2}{3}$$

cannonballs in the stack. The stack of 20 balls has four layers, and indeed

$$\binom{4+2}{3} = \frac{6 \times 5 \times 4}{6} = 20.$$

Now, imagine that the stack is fitted into a regular tetrahedron whose faces are tangent to the various balls on the outside of the stack. Also, we cut this tetrahedron into cells, with each cell enclosing a ball. The sides of the cells are the planes that are tangent to the balls and separate the balls from one another.

If we were to separate a tetrahedral stack with five layers into cells in this way, the most regular cell would be the one enclosing the cannonball that is surrounded entirely by other balls. It turns out that this cell is the *rhombic dodecahedron* shown in Figure 2.

Rhombic dodecahedra fill space just the way an infinitely large stack of cannonballs would fill space. We could cut our regular tetrahedron from this three-dimensional lattice of dodecahedra, extend the walls of some of the cells to meet the walls of the tetrahedron, and delete some others (notably at the vertices of the tetrahedron), and get the cellular decomposition of the tetrahedron shown in Figure 3.

The nice property of this decomposition of the tetrahedron is that the same shapes of cells can be used to form tetrahedra of various sizes. We name the central dodecahedral cell the *interior*

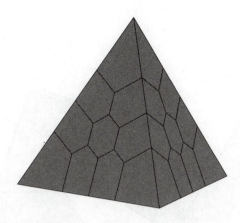

Figure 3. Satterfield's Tomb.

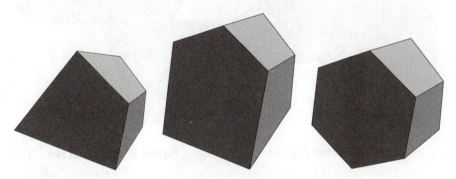

Figure 4. Vertex cell.　　Figure 5. Edge cell.　　Figure 6. Face cell.

cell and name the other cells *vertex*, *edge*, and *face* cells, as shown in Figures 4, 5, and 6.

To build a tetrahedron with L layers, we need

$$4\binom{L-2}{0} \quad \text{vertex cells,}$$

$$6\binom{L-2}{1} \quad \text{edge cells,}$$

$$4\binom{L-2}{2} \quad \text{face cells, and}$$

$$\binom{L-2}{3} \quad \text{interior cells.}$$

In particular, when $L = 4$,[1] we use 4 vertex cells, 12 edge cells, 4 face cells, and 0 interior cells.

---

[1]As usual, we define the notation $\binom{n}{k} = \frac{n!}{k!(n-k)!}$ when $0 \leq k \leq n$ and $\binom{n}{k} = 0$ otherwise.

Figure 7. Vertex cell net.   Figure 8. Edge cell net.

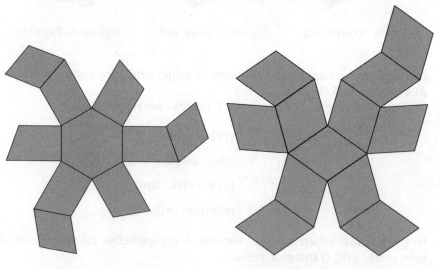

Figure 9. Face cell net.   Figure 10. Interior cell net.

One of our fondest wishes is that some company would manufacture great numbers of these cells and make them available to us and interested readers. Until this wish is fulfilled, we must satisfy ourselves by making copies of the cells from tastefully colored, stiff construction paper. To this end, we have furnished nets for the various cells in Figures 7, 8, 9, and 10.

We recommend that the reader copy these nets with a larger magnification. Then, tape a net temporarily to a construction card and put pins through the vertex points of the pattern into the card. Use a utility knife together with a steel rule to score the folds and cut out the shapes. Another method is to copy the patterns directly onto the construction paper. Some skill is required in taping the cells, and a real expert must conceal all tape inside the cell![2]

The Story of the Four Little Bears

There is a puzzle on the market that involves assembling various pieces made out of balls into a six-layer tetrahedral stack of balls. The pieces are made of marbles glued together at points. At least four of the pieces of the puzzle are shaped like little bears, as suggested in Figure 11.

Figure 11. The little bears.

It is a remarkable fact that these four little bears can be assembled in two quite different ways to form a four-layer tetrahedral stack. These two assemblages are indicated in Figures 12 and 13.

We wondered what the little bears would look like if they were made to fill the tetrahedron without leaving holes between the marble segments of the bears. Imagine the marbles enclosed in a tetrahedron, and then let the marbles expand like soap bubbles until they press against each other and the constraining walls of the tetrahedral box enclosing them. What do the cells become? And after gluing together the cells of a little bear, what would the shape

---

[2]One could also add tabs, and use glue and tape.

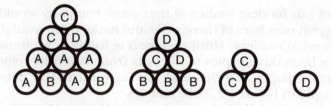

Figure 12. The little bears form a four-layer pyramid.

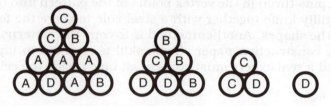

Figure 13. The little bears form a four-layer pyramid a different way.

of a bear be like? The bears interlock in a complicated way; would they pull apart when they were made of cells instead of marbles? What other shapes analogous to the bears could fill the four-layer tetrahedron? All of these questions concerning the four little bears, and our inability to imagine in detail what they would look like assembled into a tetrahedron, were the inspiration for this article.

One of us (Satterfield) used a computer to determine the various shapes of the cells: all the measurements of the line segments, the angles between them, the dihedral angles between planar faces, and so on. These detailed measurements enabled us to build and photograph a tinkertoy-like model. We also used the computer to compute the nets shown in Figures 7, 8, 9, and 10. Because of the near-death state induced by the effort involved in writing these computer programs, we refer to the structure in Figure 3 as *Satterfield's Tomb*.

We urge the reader to make a set of colored cells to use while reading the rest of this paper. Make the cells for four bears colored, say, red, blue, green, and purple. Each colored set of cells consists of one vertex cell, three edge cells, and one face cell. Our first question is, how many different animals can be made with these five cells? To answer this question, we need some notation provided by Figure 14. The figure shows a numbering of the cells as seen on the surface of the tetrahedron, cut along some of its edges, and flattened. This is the numbering we will use for the cells in the tetrahedron.

A Lifetime of Puzzles

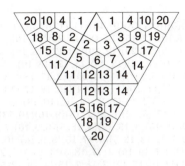

Figure 14. Cell numbering for *Satterfield's Tomb*.

Another useful device we need is the permutations of the cells induced by rotating and/or reflecting the tetrahedron. There are 24 symmetries of the tetrahedron corresponding to rotations and reflections. Such symmetries permute the four vertex cells among themselves, the twelve edge cells among themselves, and the four face cells among themselves. Table 1 lists these 24 permutations.

Armed with a notation for the cells and the permutations of the cells induced by rotation and/or reflection, we now argue that there are exactly 19 five-celled animals consisting of one vertex cell, three edge cells, and one face cell. From now on, an *animal* means a connected set of cells of these three types in these numbers.

We can assume without loss of generality that the animal has vertex cell 1. (If the vertex cell is 11, 14, or 20, we use inverses of the permutations b_1, c_1, or d_1, respectively, to change the vertex cell into 1). If 1 is the vertex cell, the face cell is either 6, 8, 9, or 16. First, consider face cell 16. There are exactly six animals with cells 1 and 16, namely, $\{1, 2, 5, 12, 16\}$, $\{1, 2, 5, 15, 16\}$, $\{1, 4, 10, 19, 16\}$, $\{1, 3, 7, 13, 16\}$, $\{1, 3, 7, 17, 16\}$, and $\{1, 4, 10, 18, 16\}$. It is easy to check that these cells are

$$\{a_i \cdot 1, a_i \cdot 2, a_i \cdot 5, a_i \cdot 12, a_i \cdot 16\}$$

for $i = 1, 2, 3, 4, 5$, and 6, respectively, so they are really all equivalent. We take the lexically smallest of these as representative. So far, we have one animal, and all animals having the vertex cell opposite its face cell are congruent to this one: $\{1, 2, 5, 12, 16\}$.

Now, we consider animals with vertex cell 1 and face cell either 6, 8, or 9. First, note that there is no loss in generality to assume that, if the face cell is one of these three, it can be assumed to be 6. If the face cell is 8, apply a_2 to change it to 6. If the face cell is 9, apply a_3 to change it to 6. Both a_2 and a_3 do not change cell 1, so

$$
\begin{array}{ll}
a_1: & (1)(2)(3)(4)(5)(6)(7)(8)(9)(10)(11)(12)(13)(14)(15)(16)(17)(18)(19)(20) \\
a_2: & (3,4)(6,8)(7,10)(12,15)(13,18)(14,20)(17,19) \\
a_3: & (2,4)(5,10)(6,9)(11,20)(12,19)(13,17)(15,18) \\
a_4: & (2,3)(5,7)(8,9)(11,14)(12,13)(15,17)(18,19) \\
a_5: & (2,3,4)(5,7,10)(6,9,8)(11,14,20)(12,17,18)(13,19,15) \\
a_6: & (2,4,3)(5,10,7)(6,8,9)(11,20,14)(12,18,17)(13,15,19) \\
b_1 a_1 = b_1: & (1,11)(2,5)(3,12)(4,15)(7,13)(9,16)(10,18) \\
b_1 a_2 = b_2: & (1,11)(2,5)(3,15)(4,12)(6,8)(7,18)(9,16)(10,13)(14,20)(17,19) \\
b_1 a_3 = b_3: & (1,20,11)(2,10,15)(3,19,12)(4,18,5)(6,9,16)(7,17,13) \\
b_1 a_4 = b_4: & (1,14,11)(2,7,12)(3,13,5)(4,17,15)(8,9,16)(10,19,18) \\
b_1 a_5 = b_5: & (1,14,20,11)(2,7,19,15)(3,17,18,5)(4,13,10,12)(6,9,16,8) \\
b_1 a_6 = b_6: & (1,20,14,11)(2,10,17,12)(3,18,7,15)(4,19,13,5)(6,8,9,16) \\
c_1 a_1 = c_1: & (1,14)(2,13)(3,7)(4,17)(5,12)(8,16)(10,19) \\
c_1 a_2 = c_2: & (1,20,14)(2,18,13)(3,10,17)(4,19,7)(5,15,12)(6,8,16) \\
c_1 a_3 = c_3: & (1,14)(2,17)(3,7)(4,13)(5,19)(6,9)(8,16)(10,12)(11,20)(15,18) \\
c_1 a_4 = c_4: & (1,11,14)(2,12,7)(3,5,13)(4,15,17)(8,16,9)(10,18,19) \\
c_1 a_5 = c_5: & (1,20,11,14)(2,19,5,17)(3,10,15,13)(4,18,12,7)(6,9,8,16) \\
c_1 a_6 = c_6: & (1,11,20,14)(2,15,19,7)(3,5,18,17)(4,12,10,13)(6,8,16,9) \\
d_1 a_1 = d_1: & (1,20)(2,18)(3,19)(4,10)(5,15)(6,16)(7,17) \\
d_1 a_2 = d_2: & (1,14,20)(2,13,18)(3,17,10)(4,7,19)(5,12,15)(6,16,8) \\
d_1 a_3 = d_3: & (1,11,20)(2,15,10)(3,12,19)(4,5,18)(6,16,9)(7,13,17) \\
d_1 a_4 = d_4: & (1,20)(2,19)(3,18)(4,10)(5,17)(6,16)(7,15)(8,9)(11,14)(12,13) \\
d_1 a_5 = d_5: & (1,11,14,20)(2,12,17,10)(3,15,7,18)(4,5,13,19)(6,16,9,8) \\
d_1 a_6 = d_6: & (1,14,11,20)(2,17,5,19)(3,13,15,10)(4,7,12,18)(6,16,8,9)
\end{array}
$$

Table 1. Permutations of the twenty cells induced by symmetries of the tetrahedron, in cycle notation. For example, $(2,3,4)$ denotes that the permutation maps face 2 to face 3, face 3 to face 4, and face 4 to face 2.

we have not violated the earlier assumption that the vertex cell is 1. Given that the vertex cell is 1 and the face cell is 6, it is fairly easy to see that, without any loss of generality, one of the three edge cells can be assumed to be cell 2. Because cell 1 has to be connected to the other cells in the animal, we must select a nonempty subset of the edge cells $\{2, 3, 4\}$. Some of these subsets contain cell 2 in the first place: $\{2\}$, $\{2, 3\}$, $\{2, 4\}$, $\{2, 3, 4\}$. Two subsets that do not contain 2 are $\{3\}$ and $\{3, 4\}$, but in these cases we can apply a_4, which does not change cell 1 or cell 6, but changes cell 3 to cell 2. Only the subset $\{4\}$ remains, but there is no connected animal that includes cells 1, 4, and 6 but excludes cells 2 and 3.

So, now we want to find all inequivalent animals that include cells 1, 2, and 6 and include two more cells from the remaining eleven edge cells 3, 4, 5, 7, 10, 12, 13, 15, 17, 18, and 19. We assume in turn that the smallest of these cells selected is 3, 4, 5, ... and see whether there is a second cell larger than it that connects the animal. It is easy to check under these assumptions that

A Lifetime of Puzzles

A_1: $\{1,2,3,4,6\}$ A_7: $\{1,2,4,6,12\}$ A_{14}: $\{1,2,6,7,12\}$
A_2: $\{1,2,3,5,6\}$ A_8: $\{1,2,4,6,13\}$ A_{15}: $\{1,2,6,7,13\}$
A_3: $\{1,2,3,6,12\}$ A_9: $\{1,2,5,6,7\}$ A_{16}: $\{1,2,6,7,17\}$
A_4: $\{1,2,4,5,6\}$ A_{10}: $\{1,2,5,6,12\}$ A_{17}: $\{1,2,6,12,13\}$
A_5: $\{1,2,4,6,7\}$ A_{11}: $\{1,2,5,6,13\}$ A_{18}: $\{1,2,3,12,15\}$
A_6: $\{1,2,4,6,10\}$ A_{12}: $\{1,2,5,6,15\}$ A_{19}: $\{1,2,6,13,17\}$
A_{13}: $\{1,2,5,12,16\}$

Table 2. Representatives of the 19 congruence classes of animals.

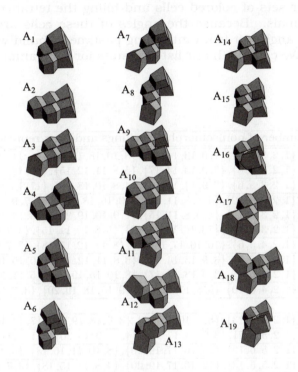

Figure 15. The nineteen animals. The little bear is A_{15}.

we can only get equivalent animals when cell 3 is selected. Then, we get equivalent pairs $\{1,2,6,3,5\}$ and $\{1,2,6,3,7\}$ (apply a_4 to the first of these to get the second) and $\{1,2,6,3,12\}$ and $\{1,2,6,3,13\}$ (again apply a_4 to see that they are congruent).

The complete list of animals is given in Table 2 and Figure 15.

Now we turn to the problem of determining which of these nineteen animals can be used to fill the tetrahedron with four congruent copies of themselves. It can be assumed without loss of generality that one of the copies is a representative A_i listed in Table 2. Then we have to apply one of the permutations $b_1, b_2, b_3, b_4, b_5,$ or b_6 to A_i to fill vertex 11, then apply one of the permutations $c_1, c_2, c_3, c_4, c_5,$ or c_6 to A_i to fill vertex 14, and finally apply one of the permutations $d_1, d_2, d_3, d_4, d_5,$ or d_6 to fill vertex 20. Such a solution might be recorded $(A_i : b_u, c_s, d_t)$. Not all of these are distinct. It might happen that $A_i = a_4 A_i$, in which case $(A_i : a_4 b_u, a_4 c_s, a_4 d_t)$ is a solution equivalent to $(A_i : b_u, c_s, d_t)$.

The reader may want to have fun looking for these solutions by making four sets of colored cells and filling the tetrahedron with colored animals. Because the angles of these cells are not the usual right angles, it is an intriguing pastime to build with these little cells! We close with our list of results for each animal.

Results

Animal	Number of nonisomorphic packings and their representatives	
A_1	2	$\{1,2,3,4,6\}, \{7,9,13,14,17\}, \{8,10,18,19,20\}, \{5,11,12,15,16\},$
		$\{1,2,3,4,6\}, \{7,9,13,14,17\}, \{5,8,11,12,15\}, \{10,16,18,19,20\}$
A_2	1	$\{1,2,3,5,6\}, \{7,9,14,17,19\}, \{4,8,10,18,20\}, \{11,12,13,15,16\}$
A_3	3	$\{1,2,3,6,12\}, \{4,5,8,11,15\}, \{13,16,18,19,20\}, \{7,9,10,14,17\},$
		$\{1,2,3,6,12\}, \{4,5,8,11,15\}, \{7,9,10,19,20\}, \{13,14,16,17,18\},$
		$\{1,2,3,6,12\}, \{13,16,18,19,20\}, \{5,8,10,11,15\}, \{4,7,9,14,17\}$
A_4	1	$\{1,2,4,5,6\}, \{10,16,17,19,20\}, \{8,11,12,15,18\}, \{3,7,9,13,14\}$
A_5	2	$\{1,2,4,6,7\}, \{3,9,13,14,17\}, \{5,8,11,12,18\}, \{10,15,16,19,20\},$
		$\{1,2,4,6,7\}, \{3,9,13,14,17\}, \{8,10,15,19,20\}, \{5,11,12,16,18\}$
A_6	1	$\{1,2,4,6,10\}, \{5,8,11,12,13\}, \{9,15,18,19,20\}, \{3,7,14,16,17\}$
A_7	0	
A_8	1	$\{1,2,4,6,13\}, \{5,8,10,11,12\}, \{3,9,18,19,20\}, \{7,14,15,16,17\}$
A_9	1	$\{1,2,5,6,7\}, \{3,9,14,17,19\}, \{4,8,10,15,20\}, \{11,12,13,16,18\}$
A_{10}	1	$\{1,2,5,6,12\}, \{13,16,17,19,20\}, \{8,10,11,15,18\}, \{3,4,7,9,14\}$
A_{11}	1	$\{1,2,5,6,13\}, \{12,16,17,19,20\}, \{4,8,11,15,18\}, \{3,7,9,10,14\}$
A_{12}	1	$\{1,2,5,6,15\}, \{9,14,17,18,19\}, \{3,4,8,10,20\}, \{7,11,12,13,16\}$
A_{13}	1	$\{1,2,5,12,16\}, \{6,13,17,19,20\}, \{9,10,11,15,18\}, \{3,4,7,8,14\}$
A_{14}	2	$\{1,2,6,7,12\}, \{3,9,10,14,17\}, \{4,5,8,18,20\}, \{11,13,15,16,19\},$
		$\{1,2,6,7,12\}, \{3,9,10,14,17\}, \{4,5,8,11,18\}, \{13,15,16,19,20\}$
A_{15}	3	$\{1,2,6,7,13\}, \{5,8,10,11,18\}, \{12,15,16,19,20\}, \{3,4,9,14,17\},$
		$\{1,2,6,7,13\}, \{5,8,10,11,18\}, \{3,4,9,19,20\}, \{12,14,15,16,17\},$
		$\{1,2,6,7,13\}, \{5,8,10,15,20\}, \{3,4,9,14,17\}, \{11,12,16,18,19\}$
A_{16}	0	
A_{17}	1	$\{1,2,6,12,13\}, \{4,5,8,10,11\}, \{3,7,9,19,20\}, \{14,15,16,17,18\}$
A_{18}	0	
A_{19}	1	$\{1,2,6,13,17\}, \{5,12,16,19,20\}, \{3,4,8,11,15\}, \{7,9,10,14,18\}$

Acknowledgments. This paper was intended to be a contribution to an earlier tribute volume to Martin Gardner, but was never completed. It is published here for the first time with the permission of Wade Satterfield and Kara Lynn Klarner.

Give a small boy a hammer and he will find that everything he encounters needs pounding.
~Abraham Kaplan

Whenever anything remotely face-like enters our field of vision, we are alerted and respond.
~H. Gombrich

Computer–Assisted Seashell Mosaics

Ken Knowlton

A typical mosaic, seen from a distance, appears to be a landscape or a portrait. But, as you approach it, you see neither leaves nor whiskers, but small separated pieces like tiles or seashells. The picture you saw from a distance fades, particularly if the pieces have interesting shapes and patterns. (In technical terms, high spatial frequencies mask lower spatial frequencies, on many levels of processing, from retinal to semantic.) The intrigue of mosaics comes largely from seeing, at a distance, more than is there.

To make a mosaic, we easily imagine starting with a sketch, painting, or photograph and replacing subareas with pieces of the same kind (*tesserae* in mosaic lingo). What I describe here is my own experience in making mosaic seashell portraits, along with the obvious questions:

1. how/whether to preprocess a picture,

2. how to fragment the picture into subareas, and

3. how to choose the tessera for each area.

My answer to the larger question of whether to use computers in these steps is obvious: as Kaplan would have expected, my shtick is computer graphics, and therefore every image that confronts me needs my kind of pounding.

The seashells that I'll talk about come in an almost-sepia range of tones, but I speak of them, and the input pictures that guide the work, as grey scale values.

Picture Enhancement

Some pictorial information is lost when turning a picture into a mosaic. The crucial question is, how do we best preserve what we want to see? An information theorist might say, "first of all, use the available shades of grey equally, i.e., level the histogram of the picture so that it has equal amounts of each shade: black, very dark, medium dark, etc." The expert might also say, "increase local contrast so that small features just slightly lighter or slightly darker than their immediate surroundings are made much lighter or much darker." (However, it is obviously difficult, though not impossible, to show detail smaller than the tesserae.)

These preprocessing methods have generally been used in picture preparation for the seashell portraits to be described, before proceeding to the main planning phase.

Planning

The following sections describe, in chronological order, five methods that I've used for planning the seashell portraits. I think of each as a successive improvement to the previous; your appraisal may differ.

(One arguably instructive method, which is not included here, is an attempt at more authentic representation: completely avoid picture enhancements and map the brightness range of available seashells as directly as possible to the dynamic range of the original picture. I tried this, and the result was so washed out that it served only as my entry in an exhibit of "Vague Art" in Phoenix and, later, as a huge Frisbee^{TM} that I flung into a New Hampshire landfill.)

Large Pixels, White Tesserae

After the starting image is enhanced, a computer graphicist's first inclination is to chop the picture regularly into "big pixels" and represent each such pixel with an object whose reflectance is roughly

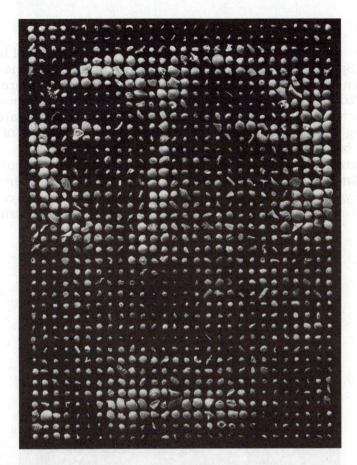

Figure 1. Square tiling with different sized shells of the same shade (Jacques Cousteau, 1987, collection San Franscisco Exploratorium).

the average value within the pixel. Figure 1 resulted from such a process, where I replaced each square tile with a light-meter calibrated seashell that reflects the called-for amount of light. These highly bleached shells, and associated teeth, vertebrae, and bits of coral, came from a small section of beach on Vieques Island; all are essentially white. Intermediate shades were achieved, in effect, by choosing the proper sizes of pieces, thus getting the right "brightness" from the percentage of occupancy (similar to the dark end of the halftone printing scale, where different sizes of white spots are regularly positioned on a black background).

The example of Figure 2 used seashells of many shades of brightness. Spacing was again according to a grid of squares. One might alternately consider a hexagonal tessellation: hexagons are more rounded and thus better approximations to seashell shapes. The regularity of either tessellation lends itself to straightforward processing methods and data structures; for similar reasons of regularity, both tessellations are common in tiled floors.

There is, however, a problem. An input picture may be cut inappropriately by any regular grid laid upon it. In a portrait, the white of an eye, or the iris or pupil, might be distributed into two or more subareas and not show clearly in the result. The same mismatch,

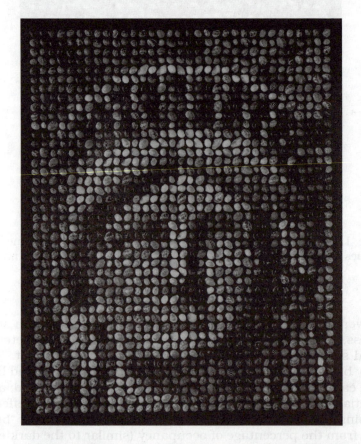

Figure 2. Square tiling with similarly sized shells of different shades (Statue of Liberty, 1991).

to a lesser degree, can happen with other facial features- mouth, eyebrows, lines, and shadows.

Fortunately, we can tweak the picture somewhat, because people tolerate modest distortion. Specifically, we can interactively distort the picture as we watch it onscreen under the proposed grid and get crucial features of the picture lined up with subareas of the grid. In practice, I have found that each corner of a portrait can be moved independently, a few percent of the picture size—stretching, shearing, and/or keystoning the whole picture—while producing no obtrusive effects. In fact, instead of interactively distorting the picture, I find it easier to interactively distort the lines of the grid, and then apply the inverse map to appropriately distort the picture to straighten the grid (imagining the grid as latching everywhere onto the picture and righting itself).

Automatic Picture-Appropriate Picture Division

With seashells and other irregular tessera, why not use freer division of the picture into subareas, making the division more appropriate for the picture and taking advantage of varied shapes? More ambitious yet, why not let a computer program determine picture fragmentation? Here's a description of one such attempt.

The approach starts with the picture tessellated by small regular hexagons. The idea is to collect these hexagons automatically into tight groups of three, four, and five, each group suggesting the size, shape, and light reflectance of a seashell to occupy the area, as indicated in Figure 3(a). Atomic operations are (1) to collect three free hexagons to form a "triangle," (2) to add a free hexagon to a triangle, thus forming a "diamond," and (3) to add a free hexagon to a diamond to form a "half-moon." Note that triangles can occur in two orientations, diamonds in three, and half-moons in six.

After executing several such operations, there remain some free hexagons and several choices of atomic operations to apply to them. I used the following heuristics, in order of priority, to pick which atomic operation to do next: (1) if a free hexagon has only one possible atomic operation to join it to other hexagons, perform that operation now; otherwise (2) favor operations near the center of the picture, and (3) make a combination whose contributing hexagons are most similar in light value, but (4) defer groupings in large monotone areas, as these will have many options.

Even with all of these considerations, lone ungroupable "orphan" hexagons can result; two such orphans already appear in Figure 3(a). Instead of complete backtracking, the program con-

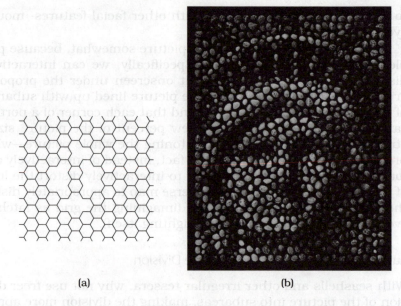

(a)　　　　　　　　　　　　　　　　(b)

Figure 3. (a) Hexagon tiling with grouping. (b) New shell mosaic resulting from that technique (Statue of Liberty, 1995).

tinues the grouping until the end and then compiles a list of the orphan hexagons that resulted. The grouping is then attempted again, with one change: the process begins by forming triangles involving all of the previous orphan hexagons. If orphan hexagons result from this grouping as well, the grouping process is repeated with additional triangles involving these orphans.

The perversity of seashells presents a final problem: they refuse to fill certain areas, particularly where the remote vertices of three diamonds and/or half-moons meet. In ad-hoc post editing, I inserted another small seashell in such a place. Conceivably, the process might better have started with smaller hexagons, to be grouped by 7s, 8s, etc., but the complexity introduced by so many patterns and orientations was daunting.

A result of this automatic picture division appears in Figure 3(b), to be compared with Figure 2's treatment of the same input picture. The mosaic in Figure 3(b) is "freer" in that one cannot easily perceive the regularity. Many consider it more attractive, possibly because there are more kinds of local patterns and various meandering paths that one may trace visually. The alert critic who observes that Figure 3(b) is hardly an improvement over Figure 2 should note that Figure 3(b) comprises only 855 seashells, as compared with the 1386 seashells in Figure 2.

Distorted Hexagonal Tessellation

Another method is to start with a complete tessellation of seashell-size regular hexagons superimposed on the input picture and interactively displace vertices, one by one, so as to make edges better fit the contours of the picture. This manual process is time-consuming, but ultimately satisfying, because space gets better filled. The method is also attractive because a small visual approximation can show the developing result, step by step. An artwork from this kind of treatment appears in Figure 4, where one can

Figure 4. Distorted hexagonal tiling done by hand (Albert Einstein, 1997, collection Donald G. McNeil, Jr.).

easily trace almost-horizontal lines across the mosaic; almost as easily, slash and backslash diagonals can be discerned throughout.

Hands–on Polygonal Tessellation

A similar, but freer, method is to start with no picture division and to interactively create a mesh by (1) creating/erasing vertices, (2) creating/erasing edges between vertices, and (3) moving vertices (as with the previous method). Figure 5 shows one result. In large smooth areas it exhibits the hexagonal tessellation seen previously; in more complex areas it contains areas more varied in size and shape.

Figure 5. Example of hands-on polygonal tessellation (Thomas Edison, 2004, collection John Caldwell).

A Lifetime of Puzzles

Other Methods and Tesserae

With a wide range of methods and tesserae, I have made other real mosaics (physical constructions) using dominoes, dice, keyboard keys, puzzle pieces, tile fragments, pottery shards, spools of thread, and jigsaw puzzle pieces; also, virtual mosaics (prints) using "alphabetic" characters (normal letters, Morse code, Braille, playing card symbols, and other alphabets); and finally, virtual mosaics made of geometric forms (triangles, squares, tangrams, pentominoes, etc.). Many of these works involve colored pictures and tesserae, paying attention, of course, to individual color components, but still considering agreement in brightness more important than agreement in red, green, or blue components.

These results can be seen on my website, http://www.KnowltonMosaics.com. (And for details about other graphics work, essays, and vita, see http://www.KenKnowlton.com.) For a wider assortment of other artists' variously perceived images, see Al Seckel's book *Masters of Deception: Escher, Dali and the Artists of Optical Illusion* (Sterling Publishing Co., New York, 2004).

The world abounds with small things that could be pieces of pictures. The challenges are to find an appropriate subject for an inviting set of pieces, or to find and use appropriate pieces to portray a worthwhile subject. Beyond such a close connection, I believe that some materials, for example seashells, become a medium in their own right, eligible for wider use.

Part V

Speak to Me

Memories and Inconsistencies

Raymond Smullyan

Here I present some memories involving (mathematical) recreations, and then some puzzles connected with Gödel's incompleteness theorem and tricking the statement-proving machine.

Some Memories and Other Things

April Fool

I first knew Martin Gardner when we were both students at the University of Chicago. Martin is a great April fooler; for example, on one April 1st an article by him appeared in *Scientific American* in which Martin claimed such fabulous things as that Leonardo da Vinci was the inventor of the flush toilet, and that, in chess, a winning move for White is pawn to king's rook 4!

I am also fond of pulling April-fool jokes. A friend might receive such a phone call:

> **Me:** Have you read the amazing article in the *New* York *Times* about Leonardo da Vinci?

> **Friend:** No, what was it?

> **Me:** There is now incontrovertible evidence that Leonardo da Vinci was really a woman!

Friend: That's fantastic!

Me: By the way, what is the date today?

Even more amusing is what might aptly be called a *meta* April-fool joke pulled by a six-year-old girl on her eight-year-old brother. On one April 1st, the brother tried an April-fool joke on his sister. The following dialogue then took place:

Sister: What's the matter with you? Today is not April Fool's!

Brother: (in amazement) It isn't?

Sister: April fool!

A Dream

When I first studied algebra in high school and heard about irrational numbers, I had the following curious dream: I was riding in a New York subway, and to my horror, it stopped at a station "$17\sqrt{2}$." I realized that I could never get to a rational numbered station again!

A Curious Puzzle

When I first studied high-school geometry, the following idea occurred to me. Imagine that you have an infinite solid plane table with a finite rod bolted perpendicular to the table. To the top of this finite rod is hinged one end of an infinite rod. The hinging allows the infinite rod to move up and down, but the curious thing is that the rod cannot possibly move down because both it and the table are solid, and therefore the rod cannot pierce the table. And so, you have the curious phenomenon of the hinged rod being supported at only one end!

Omega Inconsistency

Imagine that we are all immortal, but there is a sleeping sickness that, if you catch it, will put you to sleep forever. However, there is an antidote that will wake you up, but only for a limited time. The problem now is this: imagine that your loved one contracts the sleeping sickness today. If you give her the antidote today, she will wake up for two days and then go back to sleep forever; if you give the antidote on the next day, she will wake up for four days; and so on. In general, if you give her the antidote in n days from now, she will wake up for 2^n days. Now, you wish her to be with you awake for as many days as possible, but on any one day, if instead of giving her the antidote, you wait just one more day, you will have her for twice as long! Thus, on any one day, it is irrational to give her the antidote on that day, yet it is certainly irrational never to give her the antidote at all!

This disturbing situation might be an example of an *ω-incon-sistency* (omega inconsistency), a concept in mathematics defined as follows. Consider a mathematical system that deals with the natural numbers 0, 1, 2, . . . , n, ... It is called *inconsistent* if some sentence and its negation are both provable. The system is called *ω-inconsistent* if there is a property such that there is a proof that 0 doesn't have the property, a proof that 1 doesn't have the property, and for each natural number n, there is a proof that n doesn't have the property; yet at the same time, there is a proof that there exists a number having the property! Despite the oddness of the situation, one cannot derive a formal inconsistency from it. There are indeed consistent systems that are ω-inconsistent. The point is that a proof consists of only a *finite* sequence of sentences, so given an *infinite* sequence of sentences, even though it may be impossible that all of them are true, one cannot necessarily demonstrate this with any *finite* number of the sentences.

The situation can be nicely analogized as follows. Imagine that we are all immortal and that there are infinitely many banks in the universe: Bank 1, Bank 2, . . . , Bank n, ... You get a check saying PAYABLE AT SOME BANK, but unknown to you, it is invalid. You have taken it successively to Bank 1, Bank 2, . . . , Bank n, . . . , and at no time have any of the banks honored it. Even after you have tried billions of banks, you cannot be sure that the next bank you try will not honor it, so at no time can you prove that the check is invalid. Now, if there had been only finitely many banks in the universe, then after having tried them all and failed with each one, you would have proved that the check is invalid. But with infinitely many banks, you can never prove the check invalid, even though it is. This is an example of an ω-inconsistency.

A humorous illustration of an ω-inconsistency was given by the mathematician Paul Halmos: he defined an *ω-inconsistent mother* as a mother who says to her child: "You may not do *this*, you may not do *that*, you may not do . . . ." The child asks, "Isn't there something I can do?" The mother replies, "Yes, there is *something* you can do, but it's not *this*, nor *that*, nor . . . ."

The notion of ω-consistency will play an important role later in this article.

Some Gödelian Puzzles

I am fond of constructing puzzles that illustrate the essential ideas behind the proof of Gödel's incompleteness theorem. As many of you know, Gödel showed that, even for the most comprehensive mathematical systems of our times, there is always a sentence that is both true and unprovable in the system. He constructed a sentence that, in a certain sense, asserted its own unprovability in the system. How did he do this? The puzzles below embody the essential ideas. The first one is an updated and improved version of one that I have published previously.[1]

An Accurate Proving Machine

We consider a machine that proves various sentences constructed from the three symbols P, R, and N. By a *positive* sentence I mean any expression of one of the following two forms (where X is any expression built from those symbols):

1. PX.

2. RX.

These sentences are interpreted as follows: PX means that X is provable (by the machine, of course) and is accordingly called *true* if and only if X is provable. RX asserts that XX (the *repeat* of X) is provable, and hence is accordingly called *true* if and only if the repeat of X is provable. RX has the same meaning as PXX, and thus the two sentences are either both true or both false (depending on whether XX is provable).

By a *negative* sentence I mean one of the following two forms:

1. NPX.

2. NRX.

NPX is called *true* if and only if X is *not* provable (unprovable), and NRX is called *true* if and only if XX is unprovable.

We have an interesting loop: the machine is *self-referential* in that it proves various sentences that assert the provability or unprovability of other sentences. Now, we have given a precise definition of what it means for a sentence to be *true*. We are now given that the machine is *accurate* in the sense that every sentence

---

[1] See Raymond Smullyan's chapter "Gödelian Puzzles" in *Tribute to a Mathemagician*, A K Peters, 2004.

proved by the machine is true; it never proves any false sentences. This has several ramifications. For example, if PX is provable, so is X, because if PX is provable, it must be true, which means that X is provable. Now, suppose that X is provable; does it necessarily follow that PX must be provable? No, it does not. If X is provable, then PX is *true*, but I never said that all true sentences are provable. I said only that no false sentences are provable. As a matter of fact, there is a true sentence that is definitely *not* provable (assuming that the machine is accurate), and your problem now is to find one.

Problem 1. Find a sentence that is true, but the machine cannot prove it. (Solutions to problems are given at the end of this article.) *Hint*: Construct a sentence that asserts its own unprovability.

A Curiosity

Actually, one can construct two distinct sentences X and Y such that one of the two must be true but unprovable, but there is no way to tell which one it is!

Problem 2. Find two such sentences X and Y. (There are in fact two possible solutions.) *Hint*: Find sentences X and Y such that X asserts that Y is unprovable and Y asserts that X is provable.

Remark. I think that it was problems like the ones above that inspired Professor Melvin Fitting to once introduce me at a math lecture by saying, "I now introduce Professor Smullyan, who will prove to you that either you don't exist or he doesn't exist, but you won't know which."

A Truly Gödelian Machine

The solution to Problem 1 comes somewhat close to Gödel's argument, but Gödel never used the notion of *truth*, which was only later made precise by Alfred Tarski. What now follows comes closer to Gödel's original argument. Our present machine uses the same symbols P, R, and N, as before, as well as subscripts with the symbol 1. For any positive integer n, P_n means P followed by n subscripts of 1 (e.g., P_5 denotes P_{11111}). Our sentences are the same as before, together with sentences of the following two forms (for all n and X):

1. $P_n X$.

2. $NP_n X$.

The idea now is this: the machine proves its sentences at various *stages*, and P_nX is interpreted to mean that X is provable at stage n, while NP_nX means that X is unprovable at stage n. We are given, as part of the definition of "provable," that a sentence is provable if and only if it is provable at some stage n.

We shall refer to PX and NPX as *negations* of each other, and similarly with RX and NRX. A sentence is called *undecidable* (by the machine) if neither it nor its negation is provable. The machine is called *consistent* (or sometimes *simply consistent*) if no sentence and its negation are both provable. And now we call the system ω-*consistent* if it is simply consistent and also there is no expression X such that all the infinitely many sentences NP_1X, NP_2X, ..., NP_nX, ... are provable and at the same time PX is provable. The intuition is that PX means that X is provable at *some* stage or other, so if PX is true, there must be some n for which P_nX is true, and hence NP_nX must be false. Thus, if the machine prints only true sentences, then it cannot be ω-inconsistent. However, ω-consistency (which is the assumption that Gödel used) is much weaker than the assumption of accuracy that we used previously.

We are no longer given that the machine proves only true sentences. Instead, we are given that the machine obeys the following four conditions:

G_1: If X is provable at stage n, then P_nX is provable.

G_2: If X is unprovable at stage n, then NP_nX is provable.

G_3: If P_nX is provable for some n, then PX is provable.

G_4: If RX is provable, so is PXX, and if NRX is provable, so is $NPXX$.

The idea behind G_1 and G_2 is that, at any stage, the machine has perfect memory for what it has and has not previously proved.

From these four conditions follows a theorem:

Theorem 16.1. *G. (for Gödel) If the machine is ω-consistent, then there is an undecidable sentence. More specifically, there is a sentence G such that*

1. *if the machine is simply consistent, then G is unprovable; and*

2. *if the machine is ω-consistent, then the negation of G is also unprovable.*

The proof of Theorem G can be facilitated by first proving the following two facts:

Fact 1. If X is provable, then so is PX (assuming conditions G_1 through G_4).

Fact 2. The statement of ω-consistency implies that, if PX is provable, then so is X (again assuming conditions G_1 through G_4).

Problem 3. Prove the above facts and Theorem G, and exhibit an undecidable sentence G.

Solutions

Problem 1. One such sentence is NRNR. Recall that, for any expression X, the sentence RX is true if and only if XX is unprovable. Taking NR for X, we see that NRNR is true if and only if the repeat of NR is unprovable, but the repeat of NR is the very sentence NRNR! Thus, NRNR is true if and only if NRNR is unprovable, which means that either NRNR is true but unprovable, or false but provable. The latter alternative is ruled out by the given condition that no false sentences are provable.

Problem 2. First let us show the following. Suppose that X and Y are sentences satisfying the following two conditions:

1. X is true if and only if Y is unprovable.

2. Y is true if and only if X is provable.

Then, one of the two must be true and unprovable, but there is no way of telling which one it is! Here is the reason why. Either Y is true or it isn't. If it is true, then X is provable (as Y says); hence, X is true (because no false sentence is provable), and therefore Y is unprovable (as X says). And so, if Y is true, it is unprovable. Now, suppose that Y is false. Then Y is unprovable, so X is true (by (1)). Also, if Y is false, then X is unprovable (because Y wrongly asserts that X is provable). And so, if Y is false, then X is true but unprovable.

In summary, if Y is true, it is unprovable, and if Y is false, then X is true but unprovable. However, there is no way of telling whether Y is true or false, and hence there is no way of knowing which of X or Y is the one that is true but unprovable.

Now we need to find sentences X and Y satisfying conditions (1) and (2). One solution is to take X = NPRNPR and Y = RNPR. Another is to take X = NRPNR and Y = PNRPNR.

Problem 3.

Proof of Fact 1: Suppose that X is provable. By definition, it is provable at some stage n. By G_1, $P_n X$ is provable, so by G_3, PX is provable. $\square$

Proof of Fact 2: Suppose that PX is provable and X is unprovable. By definition, X is unprovable at any stage, so by G_2, the sentences $NP_1 X$, $NP_2 X$,... , $NP_n X$, ... are all provable. But PX is provable, contradicting ω-consistency. Hence, if the machine is ω-consistent, then the provability of PX implies the provability of X. $\square$

Proof of Theorem G: The undecidable sentence is the same as that of Problem 1—NRNR—but the proof is now different, and, I believe, more interesting.

1. Suppose that NRNR is provable. By Fact 1, so is PNRNR. On the other hand, applying the second half of G_4 with $X = NR$, because NRNR is provable, so is NPNRNR. Thus, if NRNR is provable, then PNRNR and NPNRNR are both provable, and the machine is then inconsistent. Thus, if the machine is simply inconsistent, then NRNR is unprovable.

2. Suppose that RNR (the negation of NRNR) is provable. Applying the first half of G_4 with $X = NR$, PNRNR is also provable. By Fact 2, if the machine is ω-consistent, then NRNR is provable, and the machine is then simply inconsistent. Therefore, if the machine is ω-consistent, then RNR is also unprovable, and thus NRNR is undecidable. $\square$

A Lifetime of Puzzles

A Bouquet for Gardner

Jeremiah Farrell
Thomas Rodgers

In celebration of Martin Gardner's 90th birthday, we present a bouquet of word puzzles and games based on regular solids and their graphical representations as flowers.

Puzzling Pelargoniums

Our tribute bouquet starts with several PELARGONIUMS[1]—all but two of which are red, all but two of which are yellow, and all but two of which are green. How many flowers are in our main bouquet? The reader should be able to determine the answer from just this information. This riddle is an adaptation of one of Martin Gardner's charming Snarkteasers [4, #52]. The answer will be given soon below.

---

[1] *Pelargonium* is a genus of the *Geraniacae* family of plants, with common name "geranium," notable for their brightly colored flowers. All-capital words in this article refer to flowers or flowering plants.

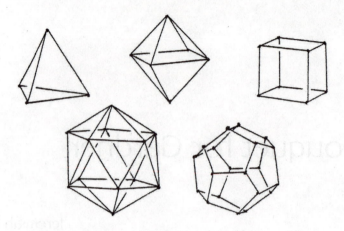

Figure 1. The five regular convex Platonic solids.

Platonic Pelargoniums

Meanwhile, there is some mathematics to be explained about our PELARGONIUMS and the hybrids that we obtain from them. We start with one of Gardner's earliest articles, "The Five Platonic Solids" [2, Chapter 1]. There he proves that the tetrahedron, the hexahedron (cube), the octahedron, the dodecahedron, and the icosahedron are the only possible regular convex solids (Figure 1).

Most of our flowers will grow from these five Platonic seeds. In his article, Gardner points out that

> the cube and octahedron are 'duals' in the sense that if the centers of all pairs of adjacent faces on one are connected by straight lines, the lines form the edges of the other. The dodecahedron and icosahedron are dually related in the same way. The tetrahedron is its own dual.

We will combine planar graphs of the solids with their duals in a special way to obtain our PELARGONIUMS. Such planar graphs are called *Schlegel diagrams*. This is a model that, as Gardner explains [8, p. 23], "is simply the distorted diagram of the solid, with its back face stretched to become the figure's outside border." In Figure 2 we draw with solid, curved lines the Schlegel diagrams of a tetrahedron, an octahedron, and an icosahedron. Superimposed with dashed lines are the Schlegel diagrams of the respective duals, the tetrahedron, the hexahedron, and the dodecahedron, respectively. The three outer dashed lines in each diagram are to be regarded as meeting in the back face.

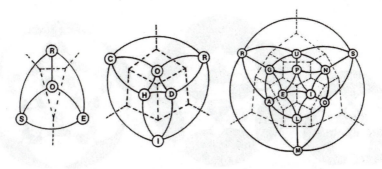

Figure 2. Schlegel graphs and their duals.

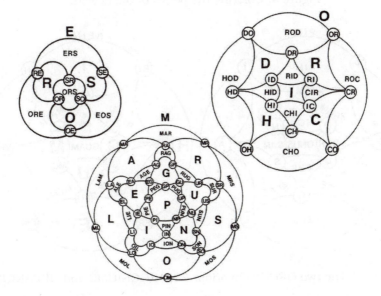

Figure 3. The three main flowers: ROSE, ORCHID, and PELARGONIUMS.

We label the nodes with the letters of ROSE, ORCHID, and PELARGONIUMS. Finally, we extend the drawings into the completed flowers by making new nodes at the intersections of the former dashed and solid arcs and then labeling each new node with the two letters on the endpoints of the former nodes. We connect two new nodes that share a common letter in their labels.[2] Figure 3 shows the resulting three flowers. These three flowers answer our opening riddle.

---

[2]In graph theory, this transformation is called the *line graph*.

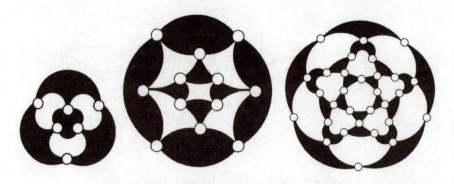

Figure 4. Coloring the petals of the flowers.

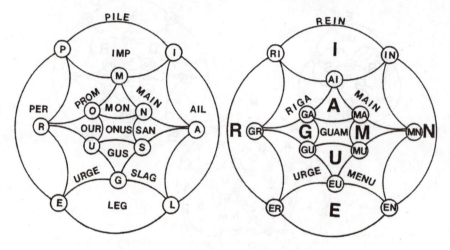

Figure 5. The two ORCHID hybrids: PELARGONIUMS and GERANIUM.

The flowers have 3-, 4-, and 5-cycles that bound new regions. Some of these cycles correspond to the former nodes of the Platonic solids, and they inherit the single letter as their label. The other cycles—the *petals* of the flower—correspond to the former regular faces of the Platonic solids. Shading these petals results in a two-coloring of the flower; see Figure 4. Each petal inherits the labels from its boundaries, and the labels have been chosen to form words that are main entries in most unabridged dictionaries or atlases. (We used the *Merriam-Webster New International Dictionary*, third edition.) We will be able to use these labels for certain puzzles and games that we have in mind on the flower graphs.

A Lifetime of Puzzles

The numbers of different flower parts are as follows:

	Nodes	3-petals	4-petals	5-petals
ROSE	6	8	0	0
ORCHID	12	8	6	0
PELARGONIUMS	30	20	0	12

Recall that the outside of each flower is also a petal.

It is possible to form hybrids by judicious relabeling of the flowers. For instance, Figure 5 illustrates the new flowers PELARGONIUMS and GERANIUM formed on ORCHID. It may be possible to obtain other hybrids using the labels WILD ROSE, VIOLET, MARIGOLD, MYRTLE, etc. Remember that the labeling should yield bona fide dictionary entries on the new names of the parts.

Puzzles on Pelargoniums

There is a famous puzzle invented in the 1850s by the Irish mathematician Sir William Rowan Hamilton that was originally played on a solid dodecahedron and that can be played on our green PELARGONIUMS grid. Gardner first described this puzzle in "The Icosian Game and the Tower of Hanoi" [1, Chapter 6]:

> [T]he basic puzzle is as follows. Start at any corner of the solid (Hamilton labeled each corner with the name of a large city), then by traveling along the edges make a complete "trip around the world," visiting each vertex once and only once, and return to the starting corner.

Today this is called finding a *Hamiltonian circuit*. The twenty 3-cycles of PELARGONIUMS are the "vertices" that must be visited in a circuit on our graph. They are each joined by a two-letter node on the flower. It is also possible to write the twenty three-letter words on tiles and try to arrange them in a chain so that abutting tiles have two letters in common. If the chain closes, you have solved the puzzle.

Gardner writes:

> On a dodecahedron with unmarked vertices, there are only two Hamiltonian circuits that are different in form, one a mirror image of the other. But if the corners are labeled, and we consider each route "different" if it passes through the 20 vertices in a different order, there are 30 separate circuits, not counting reverse runs of these same sequences. Similar Hamiltonian paths can be found on the other four Platonic solids.

One of the 30 solutions is given at the end of this article. Other informative Gardner articles about Hamiltonian circuits include "Graph Theory" [3, Chapter 10], "Knights of the Square Table" [6, Chapter 14], and "Uncrossed Knight's Tours" [4, p. 186]. John H. Conway's very interesting puzzle "A Dodecahedron-Quintomino Puzzle" [8, p. 23] can be adapted to our flower.

Gardner has often written about magic squares in his *Scientific American* columns, and we were especially interested in his report about Room squares reprinted in "The Császár Polyhedron" [10, Chapter 11]:

> A Room square is an arrangement of an even number of objects, $n + 1$, in a square array of side n. Each cell is either empty or holds exactly two different objects. In addition, each object appears exactly once in every row and column, and each (unordered) pair of objects must occur in exactly one cell.

The Australian mathematician Thomas G. Room had called this concept "A new type of magic square" in 1955, but it was later discovered that they had been in use before 1900 in scheduling bridge tournaments. We have discovered in our flowers a generalization of these squares. For instance, using the yellow ORCHID, we can form this 4×4 square from the 12 nodes:

	a	b	c	d
4	RI		HD	CO
3	CH	ID	OR	
2		OH	IC	DR
1	DO	CR		HI

This square is magic on the rows and columns in the sense that each set of three entries transposes into the word ORCHID, the magic constant. It is not a Room square because the taboo pairs IO, HR, and CD never occur together. It is instructive to locate these pairs on the ORCHID graph.

A pleasant little puzzle is possible by preparing 12 tiles with the two-letter words on them and trying to reconstruct one of the 1,152 solutions to the puzzle that look different to the eye. Two people can play the puzzle as a game by drawing a tile in turn and placing it on the grid so that no common letter occurs in any row or column. The last player to be able to play wins. To play expertly, one must heed complementary pairs of words consisting of taboo mates: RI-OH, ID-CO, IC-DO, HI-OR, CR-HD, and CH-DR. If any of these occur in the same row or column, it will be impossible to complete the trio of words in that row or column. It can also be proved that, where the blank squares in the grid intersect, we must insert complementary words. For example, the two blanks at a2 and b4 intersect at a4 and b2, where the complements RI and OH, respectively, occur.

This last property can be used in a magic trick. Let the subject find one of the possible solutions to the puzzle and then turn all the tiles face down. The subject turns over and exposes any tile, and the magician can then call out another (the complement) and is able to locate it in the grid. This is repeated until all tiles are exposed. If the grid is on a board, it may be carefully rotated before any tile is exposed, and of course the trick will still work. Interchanging pairs of rows or columns—including the blanks—can make the trick even more mysterious.

The 12 two-letter words are the edges of a three-dimensional cube with one letter at each corner. Similar square grids are possible using the edges of an n-dimensional cube. This generalization, as far as we know, has never been explored.

The 20 three-letter words in the green PELARGONIUMS flower yield another magic square. One solution with the magic constant PELARGONIUMS is the following:

MAR	OIL		SUN	PEG
PUG	MRS	ALE		ION
LIE	PUN	MOS	RAG	
	AGE	PIN	MOL	SUR
SON		RUG	PIE	LAM

The taboo pairs are MP, AN, LU, RI, GO, and ES, which lead to the complementary pairs MAR-PIN, MRS-PIE, MOS-PEG, MOL-PUG, LAM-PUN, ALE-SUN, RAG-ION, AGE-SON, SUR-LIE, and RUG-OIL.

There are 28,800 solutions to this puzzle that will look different to the eye. (Purists would not regard all of these as truly different because of certain symmetries of the grid.) Each of the remarks about the 4×4 ORCHID grid hold as well for this 5×5 PELARGONIUMS grid.

These magic squares may have applications in designing tournament schedules. For instance, suppose that we have three two-person teams that are to play one-on-one games of four types, a, b, c, and d, over four days. If the team members have initials IO, HR, and CD, respectively, our 4×4 square gives the pairings on each day.

Other Possible Pelargoniums

It is possible to extend our flower garden into higher dimensions. For example, there are six regular polytopes in four dimensions—the analogues of the Platonic solids. One is the hypercube [11, Chapter 13], but perhaps the simplest is the polytope ASTER, or the *regular simplex*, as geometers call it, whose Schlegel diagram appears in Figure 6. See also Gardner's "Tetrahedrons" [3, Chapter 19]. It has five nodes, ten lines, ten 3-cycles, and five 4-cycles. Specifically, the parts are A-REST, S-TEAR, T-SEAR, E-STAR, R-SEAT, AS-RET, AT-ERS, EA-STR, RA-SET, ST-ARE, SE-RAT, SR-TEA, TE-RAS, RT-SEA, and RE-SAT. This flower is self-dual, like the tetrahedron to which it is similar, and so a dual ASTER, interchanging nodes with 4-cycles and lines with 3-cycles, could be superimposed on the graph. When reproduced in two dimensions, the result would be a very "busy" flower!

There are of course other solids of interest that are not regular. One such infinite class is *prisms*, solids with polygonal bases and tops, with quadrilateral sides. As an example, consider the flower in Figure 7. This flower is the Schlegel diagram of a hexagonal prism. As a puzzle, we ask the reader to place the twelve letters of PELARGONIUMS in the nodes so that each of the six 4-cycles, as well as the two 6-cycles forming the base and top of the prism, transposes into words. Our solution is given at the end.

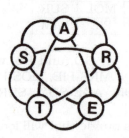

Figure 6. The ASTER.

Figure 7. The Schlegel diagram of a hexagonal prism.

Playing with Pelargoniums

The basic flower diagrams can be used for a variety of board games. One game that can be played on any of the three flowers starts by placing tokens on all the nodes. Two players alternately remove one or more tokens from any one of the cycles on the board. The player that removes the last token wins the game. This game is a Nim-type game that superficially resembles David Gale's *Chomp*. See Gardner's accounts "Nim and Tac Tix" [1, Chapter 15] and "Sim, Chomp and Race Track" [9, Chapter 9] for details. On our flower boards, however, these games are second-player wins. Can you figure out the strategy?

A more challenging game, and one we cannot predict the winner of, is *Ten Men's Morris* played on the green PELARGONIUMS flower. Each of two players has ten distinctive tokens that they alternately place on the nodes. When a player obtains a cycle of tokens, that player has formed a *Mill* and may remove one of his or her opponent's tokens. A token that is part of a Mill cannot be removed. After all their tokens have been placed, the players alternately move their tokens to empty, adjacent nodes, trying to form Mills. The game continues until one player loses by being reduced to two tokens.

All of our games and puzzles may be played strictly as word games, without using the board at all. This usually makes them immensely harder to play. In "Jam, Hot and Other Games" [5, Chapter 16], Gardner recounts a word version of tic-tac-toe by Canadian mathematician Leo Moser, who called it *Hot*. Without the symmetry of the board or grid as a guide, the games take on new life. Even our *Ten Men's Morris* game can be played as a word game. The players have a word list composed from the 3- and 5-cycles of PELARGONIUMS and begin play by drawing, in turn, ten tiles from a face-up bone pile of 30 tiles. The tiles contain the two-letter words of the nodes. When someone is able to form a word Mill with their tiles, they take a tile from their opponent's ten and place it back, face up, in the bone pile. After drawing their ten tiles, the players continue by exchanging one of the tiles in front of them with one from the bone pile that has a letter in common, still hoping to form Mills. When a player is reduced to just two tiles, they have lost the game.

Conclusion

It should be obvious that this paper could not have been written without the tremendous influence of Martin Gardner. Instead of "A Bouquet for Gardner," it would be far more proper to title this essay "A Bouquet from Gardner." We are grateful for the years of pleasure he has given us and the pleasure he continues to give us.

Solutions

Puzzling Pelargoniums. If there are n PELARGONIUMS, $n-2$ of them are red, $n-2$ of them are yellow, and $n-2$ of them are green. Thus, $n \geq (n-2) + (n-2) + (n-2) = 3n - 6$, or $n = 3$. Implicit in the puzzle is that there are only three colors, implying that, in fact, $n = 3$. (Otherwise, technically, there could be just two PELARGONIUMS, of some other color.)

Hamiltonian circuit puzzle. One solution is MAR-RAG-RUG-SUR-SUN-SON-ION-OIL-LIE-PIE-PIN-PUN-PUG-PEG-AGE-ALE-LAM-MOL-MOS-MRS.

Hexagonal prism problem. The best set of words we found is GLAMOR, SUPINE, GAIN, LUNG, PLUM, POEM, ROSE, and AIRS. (See Figure 8.)

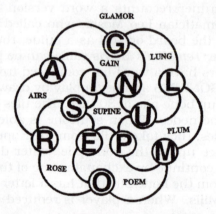

Figure 8. Solution to Figure 7.

Nim-like game. Our hint for the Nim-type game is to take advantage of the symmetry of the board, keeping in mind the complement of your opponent's play. For further insights on this kind of strategy, see Gardner's "The Game of Hex" [1, Chapter 8] and "Dodgem and Other Simple Games" [10, Chapter 12].

Bibliography

[1] Martin Gardner. *The Scientific American Book of Mathematical Puzzles and Diversions.* New York: Simon & Schuster, 1959.

[2] Martin Gardner. *The Second Scientific American Book of Mathematical Puzzles and Diversions.* New York: Simon & Schuster, 1961.

[3] Martin Gardner. *Martin Gardner's Sixth Book of Mathematical Games from Scientific American.* New York: W. H. Freeman, 1971.

[4] Martin Gardner. *The Snark Puzzle Book.* New York: Simon & Schuster, 1973.

[5] Martin Gardner. *Martin Gardner's Mathematical Carnival.* New York: Knopf, 1975.

[6] Martin Gardner. *Mathematical Magic Show.* New York: Knopf, 1977.

[7] Martin Gardner. *Mathematical Circus.* New York: Knopf, 1979.

[8] Martin Gardner. *Wheels, Life and Other Mathematical Amusements.* New York: W. H. Freeman, 1983.

[9] Martin Gardner. *Knotted Doughnuts and Other Mathematical Entertainments.* New York: W. H. Freeman, 1986.

[10] Martin Gardner. *Time Travel and Other Mathematical Bewilderments.* New York: W. H. Freeman, 1988.

[11] Martin Gardner. *The Colossal Book of Mathematics.* New York: W. W. Norton, 2001.

NetWords: A Fascinating New Pencil-Paper Game

Mamikon Mnatsakanian
Gwen Roberts
Martin Gardner

In 2004, the first author of this paper invented *NetWords*, a pencil-and-paper game that combines creative construction of words with topology. In its development phase, this game was played many times by the three authors of this paper. As they played, the three refined the rules of the game until they arrived at the version described here.

In NetWords, players write different letters of the alphabet and connect them with nonintersecting lines. They compete to find the most and longest words formed by connected letters. The result is a network densely interwoven with numerous unexpected words (hence the name NetWords).

Players will be surprised and delighted to discover what words they can form. The game can be enjoyed by both adults and children. Besides enhancing language skills, NetWords can serve as a playful introduction to the mathematical field of topology.

The structure of the game makes NetWords rich in strategy. It allows for more creativity in composing words than other popular word games, because players are free to choose letters and connect them in any direction. And the rules are still flexible, so players can change and improve them.

A Sample Game

To explain how to play NetWords, we have shown a sample game actually played by the authors. Because the game is played in turns, any number of players can participate. Two players—let's call them Ana and Bob—start with the word GAME, as shown in Figure 1.

Suppose that it's Bob's turn now. First, Bob looks for words in this mini-network, formed by four letters and three connecting lines. Words can be traced along paths of connected letters. Bob reads GAMMA by following the lines between the letters G-A-M back and forth. As you see, a letter may be included more than once in a word, and a letter may repeat itself as if it were connected to itself. It's essential to make your connections clear; thus, we suggest circling each letter as it is written.

Next, Bob may connect a pair of letters if this results in a new word. He connects E to G, forming EGG. Notice that more words, like GEM, AGE, and GAGE, have appeared. Bob claims GAGE for scoring. Bob may draw new connections between letters in the same turn, as long as each forms a word and does not cross other connection lines. Do you see another such connection that forms a word?

Finally, Bob may add a new letter and connect it to an existing letter. This is his last connection during this turn. It also should not intersect any other connection line. Bob adds N and connects it to A. He need not form a word with this connection.

Bob may search for more words in his final diagram. Surprisingly, MANNA and MANAGE emerge. Bob claims them for scoring (in addition to GAGE and GAMMA claimed earlier).

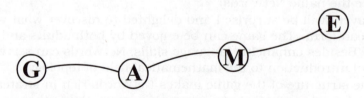

Figure 1. The beginning of a NetWords game.

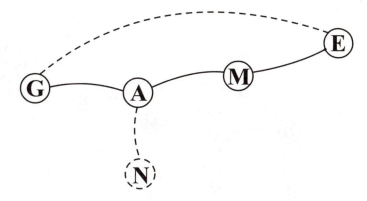

Figure 2. The end of Bob's first turn (dashed lines indicate his additions).

Bob looks for any more words, but finds none to claim. His turn ends, leaving the configuration shown in Figure 2.

Not all words may be claimed for scoring. Ana and Bob have agreed that a word must have at least three letters for connecting purposes, and at least four letters for scoring purposes. Observe that short words, like A, I, AM, and ME, are ignored, and three-letter words like EGG and MAN are not claimed for scoring.

When Ana and Bob were beginners, they used shorter words for both purposes, beginning with two-letter words. But shorter words lead to frequent connections, and thus to many smaller regions. More letters become isolated from one another, which severely limits options for longer and more interesting words. Later, they played with at least three-letter words for both purposes. Now, in this game, they decided only to score words of at least four letters. A four-letter word—being the shortest possible scoring word—counts 1 point; a five-letter word is worth 2 points; and so on. (The point value of a word is 3 less than its length.) The longer a word, the more effectively it scores. For example, a six-letter word scores the same as three four-letter words, and a nine-letter word scores the same as three five-letter words. Clearly, aiming for long words is a winning strategy.

During Bob's turn, Ana has been looking for words, too. Her turn begins with the opportunity to claim other words that she finds in the network. She claims NAME, which Bob missed. To make more words, she needs to draw more connections; each must form a word of at least three letters. Connections without intersections are easy to make early in the game. She connects E to A, claiming MEAN. She then connects E to N for AMEN.

Ana	Bob
NAME	GAMMA
MEAN	GAGE
AMEN	MANNA
MAMA	MANAGE
MANE	
GENE	
ENEMA	

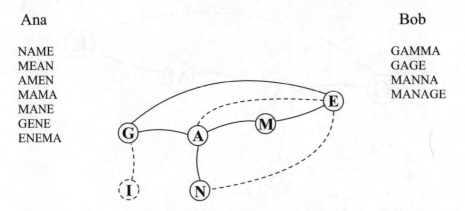

Figure 3. The end of Ana's first turn (dashed lines indicate her additions), including the words for scoring.

Bob objects to AMEN, but finds it in the dictionary. The reason for Bob's objection was that he had planned to form the word MAGMA, but can't, since it is now impossible to connect G to M without crossing other lines. All words containing these two adjacent letters will now be impossible to form in this game. A player's hopes can be dashed with a single stroke of the pen.

Ana looks for more words; she finds and claims MAMA, MANE, GENE, and ENEMA. Then, Ana adds the letter I, as shown in Figure 3, and connects it to G. She finds no other words to claim and turns the game over to Bob.

Ana realizes that she could have formed IMAGE by placing I in a different region, adjacent to the letter M, although IMAGINE would still be impossible.

At this point in the game, the network contains six letters and eight connections. Each player's words are shown on the playsheet. Ana has 8 points, and Bob has 8 points.

Bob connects I to N for NINE. He connects I to A for AGAIN. Bob adds C and connects to I. Then, he finds MAGIC, and MAGICIAN appears. Bob ends his turn.

Ana claims CINEMA. She adds T, connects it to E, claiming NINETEEN, TEENAGE, and GAMETE. Ana ends her turn.

Bob connects N to T, claiming GIANT. He connects I to T for GIGANTIC and GENETIC. Bob connects C to E for NICE and I to E for NIECE. He then adds L, which he connects to I. He claims ILLICIT and LIGAMENT. Only after his turn does Bob notice that

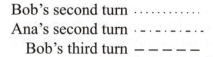

Bob's second turn ············

Ana's second turn · - · - · - · - ·

Bob's third turn — — — — —

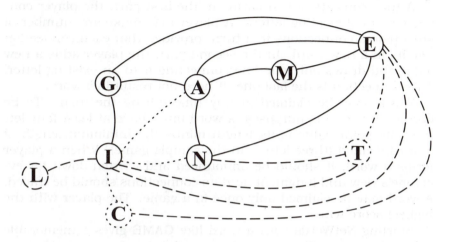

Figure 4. The end of Bob's third turn.

he has lost MANAGEMENT to Ana. Figure 4 shows the network at this point in the game.

In three more turns, Ana and Bob finished their designated 12-letter game. The game took about an hour. The final network (not shown) contained some 100 words, including ANTENNA, CIGARETTE, ELEGANT, ELEMENT, GARAGE, GARRET, INTENDED, INTELLIGENT, LENIENT, MEANDER, REMAINDER, and SILENT. The players' total scores exceeded 200 points.

We suggest that you try to reconstruct the diagram and find some of the many words not mentioned above.

Formal Rules and Guidelines

To summarize, here is a list of the formal rules of NetWords.

A letter of the alphabet may be written only once in a game. Players decide ahead on the total number of letters in the game.

A word is a sequence of connected letters with a minimum length (three letters in the sample game). Words must either be accepted by all players or listed in an agreed-upon dictionary. A letter can be repeated or duplicated in tracing a word.

Two structural rules are essential to the game's subtlety. First, a line connecting two letters must not cross any other connection lines. Second, only one connection line is allowed between two letters, and a letter cannot be connected to itself.

A turn consists of two parts. In the first part, the player connects pairs of existing letters. A player may make any number of such paired connections in a turn, provided that each connection results in a new word. In the second part, the player adds a new letter and draws only one line connecting it to an existing letter. This connection is the last one. It need not result in a word.

Words may be claimed at any time during the turn. To be claimed for scoring purposes, a word must have at least four letters. Its point value is its length minus the minimum length of a word defined (three letters in the sample game). When a player forms a word, it should be announced and written down for everyone's view and judgment, and its connections should be traced. A word may be claimed only once in a game. The player with the highest score wins.

Starting NetWords with a word like GAME gives a memorable title to that game. Players may begin with several connected letters, or they could start by simply adding and connecting one letter at a time. In this sample game, Bob could start with A and Ana could form GAME; in that case, the first claimed word may serve as a convenient title for that particular game.

After the last letter is written, each player takes a turn without adding letters to find more words to claim. A final group brainstorming session (unscored) will usually reveal a surprising number of overlooked words. Using a different color for each player helps everyone see and evaluate each individual's input.

A friendly "no-timing" approach works well. If necessary, "idle timing" can be used: if no word is claimed within a minute, the turn ends.

Proper nouns, abbreviations, and acronyms are not advised. Slang words can be used. If a player has variations of the same word (plural, roots, forms with prefixes or suffixes, adjective or adverb forms), s/he can claim only the longest. Thus, if a player who has claimed GAMESTER (for 5 points) later claims GAMESTERS, only 1 more point is added to that player's score. If a different player claimed GAMESTERS, however, that player would score the full 6 points for a nine-letter word, and the player who originally claimed GAMESTER would still earn the 5 points scored earlier (scores are never taken away). If a longer form of a word has already been claimed, shorter versions of the word may not be claimed by any player. (Thus, if GAMES were the first word claimed in a game, no one could later claim GAME.) Players should agree in advance on what constitutes a variation on a word; for example, if RENOVATE were claimed, would players then be barred from claiming NOVA because the words have the same root?

One can see many more words that could have been formed with just a few illegal crossings, such as CENTIMETER, EMERGENCE, DETERMINANT, and INTELLIGENCE. A variant that advanced players may wish to explore is to allow players to pay a penalty—such as 3 points per crossing—in order to claim words that can be formed only by moving between one or more pair of letters that are separated by lines.

Experienced players may choose words with higher numbers of letters for connecting and claiming purposes. In our sample game, a new letter could be added in any of the regions (closed or not), but not on an existing connection line. More advanced versions of NetWords may allow players to insert a new letter in a connection line between two letters, provided it forms a word (without drawing a new line). Another version may allow players to cross connection lines with a new letter inserted at the intersection, if doing so forms words in both directions. Playing a solitaire game is always a good way to hone your strategies.

In addition to a wealth of words, there are many interesting topological features embedded in the game. For example, you could figure out how to specify the network so that the game could be played over the phone. You may also observe the "triangulation" features of the final network; for instance, we can predict that the final network in a 12-letter game will contain almost (but no more than) 30 connections. Perhaps you might try to prove that three-letter words will never repeat, so they won't need to be checked for duplication. Here's a fascinating question to ponder: what network would provide the highest total score using all the letters of the alphabet?

What about playing NetWords on a torus? That would significantly enrich the connecting options. Of course, you don't have to play on a bagel! When drawing a connection, simply match the opposite edges of the playsheet, as if it had been rolled into a cylinder, and allow a line to go off one edge of the sheet and continue from the corresponding point on the opposite edge. Treat the other pair of opposite edges as being similarly connected, and you will in effect be playing on a torus. Can you think of such a topological surface that allows tracing all the words in a dictionary without intersections?

The rules given above have been refined further by the authors as they have played more games. If you do the same, you too will discover and address many nuances along the way: what constitutes a word, how to fix various mistakes or violations with penalties, how to claim a shorter word found inside a claimed word (for example, when it does not share a common root with the longer word), and so on. In all cases, the decisions should be agreed upon by all players, though always there will be some uncertainties.

If you think of rule modifications that will make NetWords more interesting or enjoyable, the authors would be pleased to hear your suggestions.

The Wizard Is Always *In*

James Randi

In my professional life I get to meet a great number of interesting folks, some famous, others not. One of the most thoroughly enjoyable persons that I've ever met and gotten to know well is Martin Gardner. I haven't the slightest idea of when I first met him, and it seems to me that I've always known him. Martin thinks that we might have had first contact at a magicians' convention somewhere. Might be.

When I first began visiting Martin, he lived in Croton-on-Hudson, New York, in his private "kingdom of Oz" at an appropriate address: 10 Euclid Avenue. I've never summoned up enough nerve to ask Martin if he chose the house for its address or for its topology, which I suspect on close examination would have proven similar to a Klein bottle. Its many rooms were jammed with columns of full filing cabinets bearing exotic labels reading, typically, "Geometry, plane, solid, 4D and up" and "Combinational Color Cubes, Magic Squares, Logic & Misc. problems." It makes one's mouth water and mind boggle.

His bookshelves boasted originals of many classics in the field of mathematics and in the art of conjuring, as well as first editions of all the L. Frank Baum "Oz" books. One section several feet long was devoted entirely to volumes on "Hollow Earth" theories, and there must have been several shelves consisting only of his own books, in several languages and various combinations. A copy machine—using that dreadful old "thermo" paper that smelled like soap and turned brown after a few months—stood humming at "ready" so that every clipping gleaned might be copied for filing under as many headings as possible. Thus, an article on the subject of telepathy could show up under that category but also in the "ESP" file, the "Rhine" file, and the "pseudoscience" file.

Martin filed numbers. If a number was shown to be a prime, it was filed under "primes" and then given its own file so that any other of its specifics might be noted. Was it the sum of cubes? Then it went into that file, as well. Any peculiarity of any sort that Martin came upon was described and preserved. I recall the advantages of this system when I was employed by International Business Machines (IBM) to work up a presentation involving logic, multiple solutions, and new ways of approaching problems. IBM was concerned with promoting their Series-370 business machines, and I asked Martin about that specific number.

"Aha!" he said (thus also inventing a book title). "The number 370 is one of only four numbers, aside from 0 and 1, that is the sum of the cubes of its own digits. What's the next highest one?" I had no answer, and felt like a fool when he told me, "The answer is obvious once you see it." (Solutions are at the end of the article.) "And if you're interested in a Spanish connection," he continued, "turn the number upside-down." I did, and IBM was happy with the results. I'm sure Martin could have gone on and on with fascinating facts about 370, or any other number I'd have cared to choose.

Martin is the most organized person I know. His tastes are simple but in keeping with his interests. Numerous Escher prints—originals, of course, bought from the artist when no one else cared—graced the walls of 10 Euclid, and a few are still on display at his retirement home in Oklahoma. A few ingenious mechanical devices occupied various shelves, and typically a table might display some puzzle that needed a solution. One such puzzle was a group of eight cards with letters spelling out "PICTURES." Told that these cards could be rearranged to form another English word of eight letters, all I could come up with was "SCRIPTURE"—leaving me short one "R." Can you find the word that I missed?

One letter of a greatly treasured stack of letters from Martin poses this partially-filled crossword puzzle with four "across" clues (see Figure 1).Can you fill in the blanks to match the clues? Compare your answers to make sure you found the intended solution.

A silver ring is the only personal adornment that I've ever seen on Martin, shaped as a tiny Möbius strip. I suppose that there are

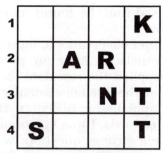

1. Intercourse
2. Noise
3. Female
4. Dirt

Figure 1. The partially filled crossword puzzle.

numerous other artifacts of this kind about, but that particular shape seems to express the man fully: it is fascinating in a direct and amusing way, has many unsuspected facets and possibilities, is simple and basic, suits him quite well, and attests to his good taste. (The original of this ring was sadly lost down a drain some years ago; the one he now wears is a replica.)

At four o'clock in the afternoon at Emerald City—pardon me, at 10 Euclid—after a long day of cerebration, a holler would come up the winding stairway (counterclockwise, two complete turns, going up) from wife Charlotte, the only other inhabitant of Oz besides the very proper cat—which one presumed was probably Cheshire, in another life. Four P.M. was "Manhattan time," and deadline or no, Martin would break away from his labors to relax. It was a ceremony carefully observed and respected by all visitors upon pain of banishment. I don't think Martin could ever drink a Martini. The possible puns that might be developed on this would be more than he could bear.

Martin impresses me in so many varied ways. Looking over just some of my correspondence with him, I note that he was never reluctant to say that he'd been wrong or ignorant about something, he always checked with me before quoting me on anything, as a proper journalist should, and his inborn sense of humor and delight with his universe was always obvious. He never stopped despairing over parapsychologists who came to silly conclusions, and in a letter sent to him by Dr. Joseph Banks Rhine of ESP fame, one that particularly amused him, he found himself referred to as "a professional denigrator."

Closing these few brief observations, let me return to that IBM symposium in San Francisco. After my presentation, I credited Martin with having supplied the raw data for the production and was pleased to see that the Systems Engineers present gave him a prolonged round of applause *in absentia*. But I was astonished when, immediately afterwards, I was surrounded by a large group of them who asked me a strange question: was Martin Gardner a real person or a composite? They found it difficult to believe that he was only one person and that he turned out such an astounding amount of material on a regular basis. They suggested that perhaps he was Isaac Asimov and John Dickinson Carr working as a team, and other combinations were also put forth.

To those folks I said, as I say to you, "Yes, there is a Martin Gardner, and he is a delight and a frustration, a wonder and a good friend to every rational mind. He's rare, generous, thoughtful, shy, valuable, and valued all in one." And he would rather I had not written any of this. But I had to.

In the last century, we had Einstein, lunar landing, instant coffee, biorhythms, black holes, and Doctor Matrix. And Martin made it all worthwhile just by being here.

Solutions

Properties of 370

The next highest number after 370 that is the sum of the cubes of its own digits is 371. A complete list of such numbers is 153, 370, 371, and 407. And turning 370 upside-down spells "OLE." ("Olé" is Spanish for "bravo.")

PICTURES anagram

The letters of "PICTURES" can be rearranged into the English word "piecrust."

Crossword

The intended words are (1) TALK, (2) BARK, (3) AUNT, and (4) SILT. Shame on you!

Part VI

Making Arrangements

Symmetric Graphs in Mathematical Puzzles

Ed Pegg Jr

Many mathematical puzzles are based on wonderfully symmetric graphs. We start with some such puzzles that the reader may wish to try to solve. Then, we describe the underlying mathematical graphs in their solutions.

Puzzle 1. Three houses (1, 2, 3) must be connected to the utilities water, gas, and electricity (W, G, E). Can all nine connections be made without any of the lines crossing?

Puzzle 2. Arrange sixteen knights on a chessboard so that each knight can attack exactly four others.

Puzzle 3. Can you arrange the ten dominoes {1–3, 2–4, 0–1, 2–3, 0–4, 1–2, 0–3, 1–4, 0–2, 3–4} in a circle so that any two neighboring dominoes do not share a number?

Puzzle 4. Write the English words {wiz, two, bet, pub, lip, sly, son, gun, zag, aye} on a sheet of paper and then connect two words if they share a letter. Can you do this with only two line crossings?

Puzzle 5. Arrange ten lines and ten points so that each line goes through exactly three points and each point is on exactly three lines.

Puzzle 6. Imagine a circular room where traveling to a point on the circular boundary teleports you to the opposite point on the circle. Using this circle, connect six points to each other (with all fifteen possible edges).

Puzzle 7. Each day, for 21 days, a five-player game of Hearts is played in the lunch room. No two people ever play in a game together more than once. What is the fewest possible number of players?

Mathematics and Solutions

Each of the puzzles above reveals some structure of a symmetric graph—in many cases, the same graph. We now describe and show these graphs, which will also reveal the solution to each puzzle. The avid puzzler is encouraged to try the puzzles above before proceeding.

First, a bit of background for the reader unfamiliar with graphs. A *graph* is a mathematical structure consisting of *vertices*, or *nodes*, and *edges*, which are point-to-point connections between pairs of vertices. Often we draw vertices as small disks and edges as straight-line connections. A *cycle* in a graph is a sequence of edges such that the first and last edges share a vertex; we call a cycle consisting of k edges a *k-cycle*. A graph is *symmetric* if there is a mapping of the graph onto itself that maps any specified vertex to any other specified vertex, and similarly there is a mapping that maps any specified edge onto any other specified edge.

Puzzle 1

It is a well-known result in graph theory, going back to 1930, that this first puzzle has no solution. As a graph, the structure is known as $K_{3,3}$, meaning that it makes all possible connections between two groups of three nodes. Here, three points (water, gas, electricity) are completely connected to three other points (houses 1, 2, 3).

To show that it is impossible to draw without crossings, consider drawing any one of the 6-cycles (shown in Figure 1). All three pairs of opposite corners remain to be connected. Either the inside or the outside must get two of these connections, and they will cause a crossing.

This graph is highly symmetric. Every edge can be either on the outer 6-cycle or one of the diagonals of this cycle. For example,

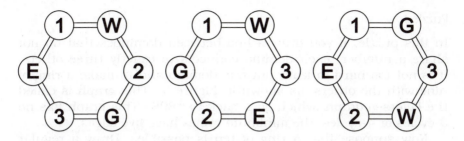

Figure 1. The three 6-cycles of the $K_{3,3}$ graph from Puzzle 1.

any edge can be chosen to be the top edge of the 6-cycle, and either endpoint of the edge can be chosen to be the top-left corner of the 6-cycle.

Puzzle 2

The solution to the knights' puzzle makes another very symmetric graph. If we draw a node for each knight and draw an edge between each pair of mutually attacking knights, we obtain the hypercube, which is shown in Figure 2.

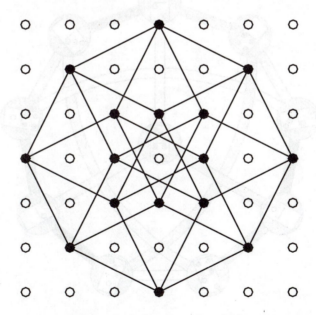

Figure 2. Solution to Puzzle 2.

Puzzle 3

In this puzzle, if you draw a line between dominoes that do not share a number, each domino connects to exactly three others. It is not too hard to leave out one domino and to make a ring of nine with the others, as shown in Figure 3. This graph is called the *Petersen graph*, which goes back to 1898. This graph has no 3-cycles or 4-cycles; the minimum cycles have five edges.

Now suppose that a ring of ten is possible. Draw a regular decagon. Five more connections are necessary. Not all of them can connect to the opposite corner of the decagon. At least one connection must be to a corner that is four corners away. Now the neighboring corners cannot be connected without making a 3-cycle or a 4-cycle. Therefore a ring of ten is not possible. In other words, the Petersen graph does not have a *Hamiltonian cycle*—a cycle visiting every vertex exactly once.

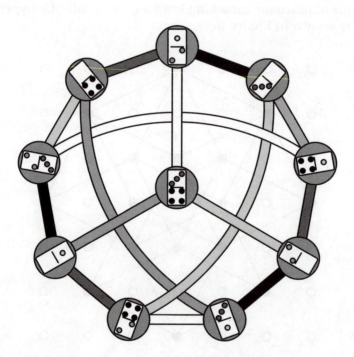

Figure 3. Petersen graph with ten dominoes.

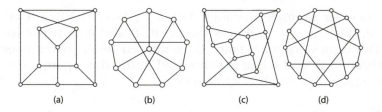

(a)   (b)   (c)   (d)

Figure 4. The Petersen graph, (a) rearranged so that only two lines cross and (b) in its traditional orientation, and the Heawood graph, (c) rearranged so that only three lines cross and (d) in its traditional orientation.

Puzzle 4

This word puzzle also gives a Petersen graph. It is possible to arrange the points and connections so that there are only two crossings, as shown in Figure 4 on the left. This drawing still has a reflectional symmetry, though it lacks the other symmetries of the Petersen graph.

As another example, the symmetric graph called the *Heawood graph*, shown above on the right, can be drawn so that it has only three crossings. All of the three-crossing embeddings that I have found are asymmetric. The crossing numbers of larger symmetric graphs are unknown.

Puzzle 5

In this puzzle, the ten points can be replaced with the ten dominoes that we used for the Petersen graph in Puzzle 3, producing the *Desargues configuration* shown in Figure 5. Consider one of

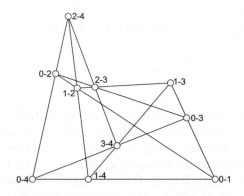

Figure 5. The Desargues configuration.

the lines, for example, through 1–2, 2–3, 1–3. The line itself can be identified by the numbers *not* on it, 0 and 4 in this case. In the Desargues graph, a point is connected to the three lines going through it, and a line is connected to the three points that are on it. Again, the graph is symmetric.

Puzzle 6

This puzzle uses the *projective plane*—the topological space caused by the circle of teleportation—to introduce the Petersen graph again. If you solved the puzzle, you might have a sketch like the one in Figure 6 on the left. If you look at the ten regions that the circle has been divided into and create the graph's *dual*—draw a vertex for each region and draw edges to represent adjacent regions—as shown in Figure 6 on the right, you will see that it is the Petersen graph again.

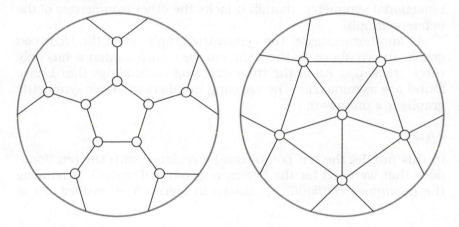

Figure 6. The solution to Puzzle 6 (left) and its dual, the Petersen graph (right).

Puzzle 7

The best solution requires only 21 players, with every player meeting every other exactly once. The first day is shown in bold; for day n, rotate the bold drawing by n notches. In the end, we obtain the *complete graph* of all possible edges K_{21}, which is symmetric. As shown on the right, a daily *6-player* game can last *31 days* with 31 players, again with every pair of players playing together exactly once.

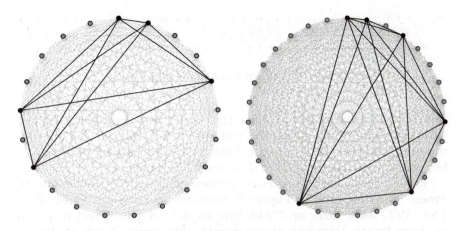

Figure 7. The complete graphs K_{21} (left) and K_{31} (right).

Drawing Symmetric Graphs

Symmetric graphs are fascinating, but I have not had much luck in finding pictures of them. For many, I had to create the pictures on my own. For example, Figure 8 is one picture that I made of the *Hoffman-Singleton graph*.

This particular graph can be constructed with triplets. Pick three numbers from $\{1, 2, 3, 4, 5, 6, 7\}$. There are 35 ways to do this, from 123 to 567. Call each of these a *triad*. Triplets 136, 613,

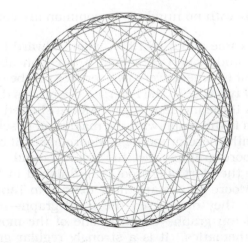

Figure 8. The Hoffman-Singleton graph.

Figure 9. Fifteen Fano planes (fanos).

361, 631, 316, and 163 all have the same three numbers and are thus the same triad, which we write in sorted order as 136. In my maa.org column [14], I present these as a set of fifty cards that can be printed out. So far, I have described 35 cards.

Fifteen more cards are *Fano planes*, which I will abbreviate to *fanos*. The first fano in Figure 9 contains seven triads: 136, 235, 145, 127, 347, 567, and 246 (the circle). If you select any two of these fanos, they will share exactly one triad. Each of the 35 possible triads is found in exactly three Fano planes.

That is enough to make the graph. Every fano is connected to seven triads. Every triad is connected to three fanos and four triads. Every card is connected to exactly seven other cards. For any two cards that are not connected, there is exactly one card connected to both. Here are a few examples: 123 and 234 are both connected to 567; 123 and 345 are both connected to the 11th fano; 123 and the last fano are both connected to 457. The graph is controlled by the following connection rules:

1. Two fanos are never connected.

2. If a fano contains a triad, they are connected.

3. Two triads with no numbers in common are connected.

This graph came about in 1960, when Edward F. Moore asked Alan Hoffman and Robert Singleton a question about graphs of diameter 2, that is, where every two vertices can be connected by a path of at most two edges. The Petersen Graph has diameter 2, and each node has three incident edges. Moore showed them that the Petersen graph is the graph with the maximum possible number of vertices and with these properties. Moore asked if other graphs—now called Moore graphs—met his upper bound. Hoffman and Singleton gave the start of a proof [12], finished in 1971, that the only possible Moore graphs are those defined in Table 1.

In addition, they discovered the third graph—now called the Hoffman-Singleton graph—which is one of the most remarkable objects in mathematics. It is a strongly regular graph, an integral graph with graph spectrum $(-3)^{21}2^{28}7^1$, the unique (7, 5)-cage

A Lifetime of Puzzles

Name	Degree Δ (edges per vertex)	Diameter d	Vertices v
Odd n-gon	2	$(n-1)/2$	N
Petersen Graph	3	2	10
Hoffman-Singleton Graph	7	2	50
???	57	2	3250

Table 1. The possible Moore graphs.

graph, and a symmetric graph. Whether the fourth graph exists has been a famous unsolved problem since 1960.

From the Hoffman-Singleton graph, we can produce another remarkable symmetric graph, named after the famous geometer H. S. M. Coxeter. Select 15 cards from the Hoffman-Singleton deck that are disconnected from each other. For example, you could select all fifteen fanos or all of the triads containing a 1. From the remaining 35 cards, discard the seven cards that are connected to one of the 15 cards. The remaining 28 cards make the Coxeter graph.

Another way to make the Coxeter graph is as follows. Form all 35 possible triads with the consonants D, L, N, P, R, S, and T. Add vowels to create words, then delete this fano of words: *plus, surd, land, pirn, nest, dept, terl*. The resulting object, shown in Figure 10, is the Coxeter graph. The words *salt* and *pond* do not share

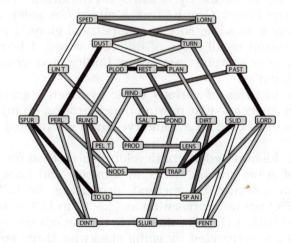

Figure 10. The Coxeter graph.

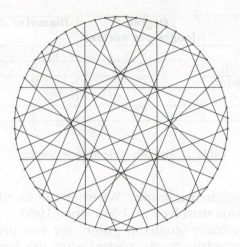

Figure 11. F64.

consonants, so they are connected. Words *lint* and *trap* share a *t*, so they are not connected.

From any word, you can reach any other word within four steps, so the Coxeter graph has a diameter of 4. H. S. M. Coxeter was a great lover of symmetry—he liked this graph so much that he wrote a paper about it called simply "My Graph" [7].

When I first started looking at symmetric graphs, I learned of the Foster Census, which is a listing of all of the cubic symmetric graphs up to 768 vertices. Figure 11 is an example of one of them, F64, found for me by Guenter Stertenbrink. For almost a year, F90—the Foster Graph—was a minor obsession of mine. I wanted to *see* it in some sort of symmetrical glory. I moved dots and lines around in all sorts of different ways. I learned many tricks—computer searches, symmetry finding, line crossings, and Hamiltonian paths.

Here are pictures of various cubic symmetric graphs. F56A was an early success—just moving nodes around on my computer screen, I got clobbered by symmetry—and I got hooked on finding more.

In 1965, Joshua Lederberg developed a method for describing hundreds of cubic graphs. H. S. M. Coxeter and Roberto Frucht subsequently modified his method, producing the *LCF notation*. The cube, F8, can be represented as $[3, -3]^4$ in LCF notation. The graph starts with a Hamiltonian cycle on the outside. After that, the vertices are connected by going clockwise three spaces, then counterclockwise three spaces—do this four times. F14 is $[5, -5]^7$,

A Lifetime of Puzzles

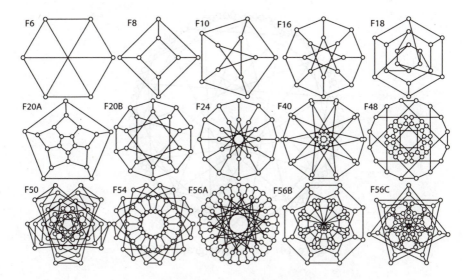

Figure 12. Various cubic symmetric graphs.

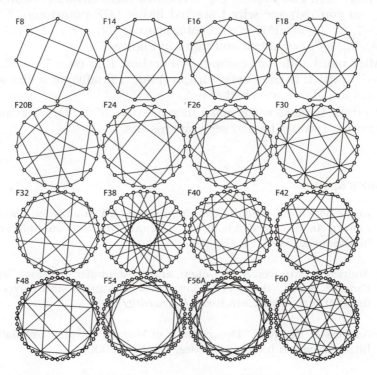

Figure 13. Cubic symmetric groups based on LCF notation.

Figure 14. The Foster graph, F90.

F16 is $[5, -5]^8$, F38 is $[15, -15]^{19}$. The raw form of F38 is somewhat long, so you can see why I started liking LCF notations. F18 is $[5, 7, -7, 7, -7, -5]^3$. F60 is a bit complicated: $[12, -17, -12, 25, 17, -26, -9, 9, -25, 26]^6$. See these and more in Figure 13.

After many different computer searches, I hit on $[17, -9, 37, -37, 9, -17]^{15}$, which is the Foster Graph (F90). I finally got a chance to *see* it. I am glad I searched.

A puzzle to close with: make nice pictures of the larger symmetric graphs. I would love to see your results.

Bibliography

[1] Roger Beresford. "Regular Graph Synthesis & Visualization." *Wolfram Library Archive*. Available at http://library.wolfram.com/infocenter/MathSource/5176/, 2004.

[2] Andries E. Brouwer, A. M. Cohen, and A. Neumaier. *Distance Regular Graphs*. Berlin: Springer-Verlag, 1989. (Additions and corrections listed at http://www.win.tue.nl/~aeb/drg/.)

[3] Andries E. Brouwer. "Descriptions of Various Graphs." Available at http://www.win.tue.nl/~aeb/drg/graphs/, 2006.

[4] Andries E. Brouwer. "Hoffman-Singleton Graph." Available at http://www.win.tue.nl/~aeb/drg/graphs/Hoffman-Singleton.html, 2004.

[5] Francesc Commellas and Charles deLorme. "The (Degree,Diameter) Problem for Graphs." Available at http://www-mat.upc.es/grup_de_grafs/grafs/taula_delta_d.html, 2006.

[6] H. S. M. Coxeter, R. Frucht, and D. Powers. *Zero-Symmetric Graphs.* New York: Academic Press, 1981.

[7] H. S. M. Coxeter. "My Graph." *Proc. London Math. Soc.* 3:46 (1983), 117–136.

[8] H. S. M. Coxeter. *The Beauty of Geometry.* New York: Dover Publications, Inc., 1968.

[9] Geoff Exoo. "The Hoffman-Singleton Graph." *Miscellaneous Topics in Combinatorics.* Available at http://ginger.indstate.edu/ge/Graphs/HOFFSING/index.html, 2006.

[10] Chris Godsil. "Problems in Algebraic Combinatorics." *The Electronic Journal of Combinatorics* 2 (1995), R8 (http://www.combinatorics .org/Volume_2/volume2.html#R8).

[11] Chris Godsil and Gordon Royle. *Algebraic Graph Theory.* New York: Springer, 2001.

[12] Alan J. Hoffman and R. R. Singleton. "On Moore Graphs with Diameters 2 and 3." *IBM Journal of Research and Development* 4:5 (1960), 497 (http://www.research.ibm.com/journal/rd/045/ibmrd0405H.pdf).

[13] D. A. Holton and J. Sheehan. *The Petersen Graph.* Cambridge, UK: Cambridge University Press, 1992.

[14] Ed Pegg Jr. "The Hoffman-Singleton Game." *MAA Online Math Games* (November 1, 2004). Available at http://www.maa.org/editorial/mathgames/mathgames_11_01_04.html.

[15] Gordon Royle. "Cages of Higher Valency." Available at http://www.csse.uwa.edu.au/~gordon/cages/allcages.html, 1997.

[16] Gordon Royle. *Cubic Symmetric Graphs (The Foster Census).* Available at http://www.cs.uwa.edu.au/~gordon/remote/foster/, 2001. (Originally published in 1934 by Ronald M Foster.)

[17] Eric W. Weisstein. "Cage Graph, Configuration, Coxeter Graph, Desargues Graph, Distance-Regular Graph, Generalized Petersen Graph, Graph Diameter, Girth, Hoffman-Singleton Graph, Integral Graph, Moore Graph, Petersen Graph, Sylvester Graph, Symmetric Graph, Vertex Degree." *MathWorld—A Wolfram Web Resource.* Available at http://mathworld.wolfram.com/, 2007.

[18] D. West. *Introduction to Graph Theory*, Second Edition. Englewood Cliffs, NJ: Prentice Hall, 2001.

[19] Stephen Wolfram. *A New Kind of Science*. Champaign, IL: Wolfram Media, 2002.

Martin Gardner and Ticktacktoe

Solomon W. Golomb

On at least three occasions (March 1957, August 1971, and April 1979), Martin Gardner devoted his "Mathematical Games" column in *Scientific American* to *ticktacktoe* (as he spells it—it is also spelled *tic-tac-toe*, and the game is known in the U.K. as *noughts and crosses*) [2–4].

This article is my personal collection of the most interesting, curious, and remarkable facts that I know about ticktacktoe and its generalizations, especially to the hypercube of side n in k dimensions (the "n^k board" discussed at length by Golomb and Hales [5]). Familiar ticktacktoe is the case of side $n = 3$ in $k = 2$ dimensions.

Ticktacktoe and Magic Squares

Consider the following game. The numbers from 1 to 9 are written on individual cards, which are placed face up on a table. Two players alternately select cards (one on each turn) from the table, to hold in their hands. The first person to hold three cards that sum to 15 is the winner. If all nine cards have been selected and neither player has a subset of three cards summing to 15, the game is a draw.

2	9	4
7	5	3
6	1	8

Figure 1. Ticktacktoe as a game on a magic square.

While it may not be immediately obvious, this game is mathematically identical to ticktacktoe. To see this, look at the magic square in Figure 1. Each row, column, and principal diagonal sums to 15. In fact, three of the numbers from 1 to 9 sum to 15 *if and only if* they lie along a winning ticktacktoe path.

Winning Paths on the n^k Board

Generalized or *hypercube* ticktacktoe is played on a "board" of a k-dimensional hypercube with "side" n, which has n^k cells. A *winning path* is any set of n cells that lie on a straight line, including diagonal lines. The number of these winning paths is $\frac{1}{2}[(n+2)^k - n^k]$. Formally, this is shown as follows. A cell is located by a *position vector* $\alpha = (a_1, a_2, \ldots, a_n)$, which specifies the k coordinates of the cell in k-dimensional space. (Each a_i is a number from 1 to n.) A *winning path* P is a sequence of n cells, $P = \langle \alpha_1, \alpha_2, \ldots, \alpha_n \rangle$, such that if we look at the ith coordinates of $\alpha_1, \alpha_2, \ldots, \alpha_n$, we see either $(1, 2, \ldots, n)$, or $(n, n-1, \ldots, 2, 1)$, or $(c, c, \ldots, c)$ for some constant value c between 1 and n. There are thus $n + 2$ possibilities for each of the k coordinates, for a total of $(n+2)^k$; however, if all k coordinates remain constant, P is not a path but merely an n-fold repetition of a single cell. Because each of the k coordinates can remain constant in each of n ways, we must subtract n^k from $(n+2)^k$ to get the number of *directed* winning paths. If we regard $\langle \alpha_n, \alpha_{n-1}, \ldots, \alpha_2, \alpha_1 \rangle$ as the same path as $\langle \alpha_1, \alpha_2, \ldots, \alpha_{n-1}, \alpha_n \rangle$, but merely traversed in the opposite direction, then the number of (undirected) winning paths is $\frac{1}{2}[(n+2)^k - n^k]$.

There is an intuitive, geometric way to interpret this formula. Embed the n^k hypercube centrally in an $(n+2)^k$ hypercube, so that there is a (hyper)shell of unit thickness around the n^k "board." When extended, each winning path on the n^k board terminates in exactly two "shell cells," and each shell cell is on an extension of

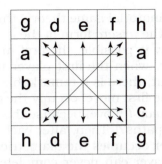

Figure 2. Each of the eight winning paths on the 3×3 board terminates in two shell cells of the embedding 5×5 board.

exactly one path. Hence, the number of paths is half the number of shell cells: $\frac{1}{2}((n + 2)^k - n^k)$.

This geometric interpretation is illustrated in Figure 2 for the familiar $n = 3$, $k = 2$ ticktacktoe board.

Ticktacktoe in Three Dimensions

To a mathematician, two-dimensional ticktacktoe is quite uninteresting. If $n = 2$, i.e., on the 2×2 board, the first player *must* win on her second turn. The 3×3 board, with best play on both sides, will always be a draw. In general, because the first move cannot be a disadvantage, the first player tries to win and the second player tries to draw. The drawing strategy for the second player on the 3×3 board is easily learned. The longer the side of the board, the easier it is for the second player to draw. (We return to this in the next section.)

This leads us to look at three-dimensional ticktacktoe ($k = 3$). On the $3 \times 3 \times 3$ "board," the central cell is so powerful that, if it is used as the opening move of the first player, a win quickly ensues. If the first player does not occupy the central cell on the first move, the second player can occupy it and force a win within several more moves. In fact, there is no way to fill the $3 \times 3 \times 3$ "board" with 14 *crosses* (Xs) and 13 *noughts* (Os) without completing one (or more) winning paths, so the game *cannot* be a draw, and with no restrictions (and with best play on both sides) it must be a win for the first player.

This leads us to an amazing result. Suppose that the rules are modified to state that the first player to complete a ticktacktoe path is the *loser*. (In combinatorial game theory, this reversal of the rule for winning is called the *misère* form of the game.) On the $3 \times 3 \times 3$ board, where no draw is possible, the first move would now seem to be a *disadvantage*, when the first player to complete three-in-a-row is the *loser*. But, there is a remarkable winning strategy for the first player: on the first move, occupy the center cell! (But, you object, isn't that the *worst* move, because the center cell lies on 13 different ticktacktoe paths?) Then, after each move by the second player, the first player occupies the diametrically opposite cell! In this way, the first player can never complete a path *through* the center (it's already blocked), and any path completed *not* through the center by the first player will be the mirror image of one just completed by the second player! So the first player cannot lose; thus, because no draws are possible on the $3 \times 3 \times 3$ "board," he must win!

The same winning strategy for misère ticktacktoe works for the first player on any n^k "board" where n is odd and no draw is possible, for $n > 1$.

Now back to the normal game, where the first player to complete a path *wins*. The *playable* three-dimensional ticktacktoe game is $4 \times 4 \times 4$, and several commercial versions of this game have been sold. There are $4^3 = 64$ cells, and $\frac{1}{2}(6^3 - 4^3) = 76$ winning paths. A number of attempts were made to "solve" this game. Finally, O. Patashnik, in 1980 [7], showed that with perfect play on both sides, the first player can win. (The exhaustive computer analysis used to show this fact does not constitute a "winning strategy" that can be easily *learned*.)

Draws by the Pairing Strategy

On the n^k "board," if the number of cells (n^k) is at least twice as large as the number of paths ($\frac{1}{2}[(n+2)^k - n^k]$), it *may* be possible for the second player to force a draw by following a *pairing strategy*: Dedicate two cells (exclusively) to each path, so that if the first player occupies one of these cells, the second player will occupy the other. (If the first player occupies an undedicated cell, or a cell paired with one already occupied by the second player on a previous "free move," the second player is free to occupy any vacant cell.)

When $k = 2$, the smallest board with at least twice as many cells as paths is the 5×5, which has 25 cells and $(49 - 25)/2 = 12$ paths. A pairing strategy for this board is shown in Figure 3.

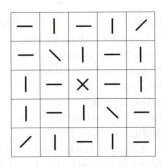

Figure 3. On this 5×5 board, each row has two dedicated cells shown by horizontal lines; each column has two dedicated cells shown by vertical lines; each diagonal has two dedicated cells shown by appropriately slanted lines; and the central cell (containing an X) is undesignated.

The smallest "board" with $k = 3$ to have at least twice as many cells as paths has $n = 8$, and a pairing strategy is known (see [5]).

In general, "at least twice as many cells as paths" is equivalent to $n^k \geq (n+2)^k - n^k$, which can be rewritten as $2n^k \geq (n+2)^k$ or $2 \geq (1 + \frac{2}{n})^k$.

Hales and Jewett [6] conjecture that whenever this inequality holds, the second player can force a draw. This conjecture has now been proved for all "large" k, e.g., for $k > 100$ (see [1]). The draw is not by a pairing strategy, however, and it remains open whether a pairing exists whenever the inequality permits. (It has been shown for the first few values of k only: $k \leq 5$.) It is reasonable to believe that when a pairing strategy exists on the n^k "board," such a strategy will exist for all *larger* values of n (with the same k) and for all *smaller* values of k (with the same n). For partial results on this, see [5].

For a given dimension k, let n_k be the smallest "boardside" n for which the number of cells is at least twice the number of winning ticktacktoe paths. That is, n_k is the smallest n for which $2 \geq (1 + \frac{2}{n})^k$. Solving for n gives $n_k = \lceil 2/(2^{1/k} - 1) \rceil$, where "$\lceil x \rceil$" means the smallest whole number greater than or equal to the real number x.

I had observed that this exponential (exact) expression for n_k seemed to give the much simpler-looking *linear* expression $n_k = \lfloor 2k/\ln 2 \rfloor$, where $\lfloor x \rfloor$ denotes the largest whole number less than or equal to the real number x, and $\ln 2 \approx 0.69315\ldots$ is the natural logarithm of 2. I tested the conjecture $n_k = \lceil 2/(2^{\frac{1}{k}} - 1) \rceil \stackrel{?}{=} \lfloor 2k/\ln 2 \rfloor$ for the first several thousand values of k, and it always held. In fact, it holds for the first *777 trillion* values of k, but it is *false*! The first counterexample occurs at $k = 777{,}451{,}915{,}729{,}368$, where the true n_k is $2{,}243{,}252{,}046{,}704{,}767$, but $\lfloor \frac{2k}{\ln 2} \rfloor$ is "only" $2{,}243{,}252{,}046{,}704{,}766$. That is, if you happened to be playing "hypercube ticktacktoe" in $k = 777{,}451{,}915{,}729{,}368$ dimensions, and the side of the board was "only" $n = 2{,}243{,}252{,}046{,}704{,}766$ (i.e., in excess of two quadrillion), the second player would not have quite enough cells (even with n^k cells—are there that many atoms in the universe?) to hope to pair them so as to dedicate two cells to every ticktacktoe path! These values for the smallest counterexample k, and the corresponding n_k, were recently found by Joe Buhler. They correct the erroneous values (due to a faulty multiple-precision computer program) given in [5], where $k = 6{,}847{,}196{,}937$, too small by five orders of magnitude, was claimed as the first counterexample. Candidates for possible counterexamples occur at the "continued-fraction convergents" of $\frac{2}{\ln 2}$, with the first actual counterexample appearing at the 36th convergent. The first six counterexamples occur at the convergents numbered 36, 40, 42, 58, 78, and 90, suggesting that there are probably infinitely many counterexamples, albeit quite sparsely distributed among the integer values of k.

The number of hypercells in the smallest "board" where $n_k \neq \lfloor \frac{2k}{\ln 2} \rfloor$, namely $(n_k)^k$ for $k = 777{,}451{,}915{,}729{,}368$, is my candidate for the largest integer that occurs as the *specific* solution to a problem that arose "naturally," i.e., not constructed specifically to lead to a huge number, and that is known *explicitly* and not merely as an upper or lower bound on some answer. (Of course, even if this is a current record, all records are made to be broken; and this record is sensitive to how the rules are formulated.)

When Are Draws Possible?

If no draw is possible on the n^k board, meaning that no matter how you fill in the board with Xs and Os one player has won, then no draw will be possible for larger k with the same n, and no draw will be possible for smaller n with the same k. Thus, for each k, there is a largest n for which no draws are possible, and this value of n is a monotonic nondecreasing function of k.

For $k = 1$, no draw is possible when $n = 1$, but a draw is forced for $n \geq 2$. For $k = 2$, no draw is possible when $n = 2$, but one can be obtained by the second player when $n \geq 3$. When $k = 3$, as mentioned earlier, no draw is possible when $n = 3$, but a draw *can* occur when $n = 4$ (though not with best play by the first player).

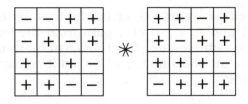

Figure 4. Two 4×4 boards whose 4^4 tensor product has no "solid" ticktacktoe paths.

For $n \geq 5$, it is believed that the second player can obtain a draw by best play, and for $n \geq 8$, the second player can do so by a pairing strategy.

Thus, for each of $k = 1$, 2, and 3, the largest n for which no draw is possible on the n^k board occurs at $n = k$ (on the 1^1, the 2^2, and the 3^3 "boards"). Is this also true for the case $n = k = 4$?

Some 45 years ago, A. W. Hales showed that a draw *can* be constructed on the 4^4 board, by the following ingenious construction (see [5]). We take the *tensor product* of the two 4×4 boards shown in Figure 4. The concept of a tensor product is illustrated in Figure 5 with two 2×2 arrays. The 2×2 arrays on the right side in Figure 5 are first stacked horizontally, to get two $2 \times 2 \times 2$ "cubic" arrays, and these are then stacked "vertically" (this may tax your four-dimensional visualization ability!) to get one 2^4 array.

In Figure 4, the *left* 4×4 array has exactly two +s and two −s on each ticktacktoe path (a nonzero *even* number of each), while the *right* 4×4 array has exactly three +s and one − on each ticktacktoe path (an *odd* number of each). In the tensor product, each of the 520 ticktacktoe paths will be one of the following types: (a) a path from one of the two 4×4 squares (hence nonsolid +s or −s), (b) the *negative* of a path from one of the two 4×4 squares (again

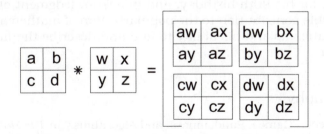

Figure 5. The tensor product of two 2^2 arrays is a 2^4 array.

nonsolid with either +s or −s), or (c) the term-by-term "product" (where + = +1 and − = −1) of a path from the left square with a path from the right square; but this term-by-term product will necessarily have an *odd* number of −s, so it cannot have all four entries the same!

Phase Transitions in the (n, k) Plane

For each value of the dimension $k \geq 1$, we can consider, with increasing n, the following situations:

(a) The first player *must* win.

(b) No draws are possible, and while a win is not forced, the first player *should* win.

(c) The first player, although draws are possible, *can* win with skillful play.

(d) The second player *can* draw with skillful play, although there is no pairing strategy to force a draw.

(e) The second player can draw by use of a pairing strategy.

Each of these situations corresponds to a region (among the lattice points in the first quadrant) of the (n, k) plane. The boundaries between these regions may be thought of as *phase transitions* (as in physics and chemistry). In a few cases, the precise boundaries are known; but more generally, qualitative statements can be made about the shapes of these regions, such as *row convex*, or *column convex*, or possibly both of these. This topic is described (along with several other things) in more detail in [5], which also has a more complete list of references.

Acknowledgment and Dedication. This article is dedicated to Martin Gardner, for his 90th birthday, and in acknowledgment of his immeasurable contribution to the popularization of mathematics and mathematicians and his ability to find and describe the *fun* in doing math.

Bibliography

[1] J. Beck. "Games, Randomness, and Algorithms." In *The Mathematics of Paul Erdős*, Vol. I, edited by R. L. Graham and J. Nešetřil, pp. 280–310. New York: Springer, 1997.

A Lifetime of Puzzles

[2] Martin Gardner. "Tichtacktoe." Chapter 4 in *The Scientific American Book of Mathematical Puzzles and Diversions.* New York: Simon and Schuster, 1959.

[3] Martin Gardner. "Tichtacktoe Games." Chapter 9 in *Wheels, Life and Other Mathematical Amusements.* New York: W. H. Freeman, 1983.

[4] Martin Gardner. "Generalized Tichtacktoe." Chapter 17 in *Fractal Music, Hypercards and More.* New York: W. H. Freeman, 1991.

[5] Solomon W. Golomb and Alfred W. Hales. "Hypercube Tic-Tac-Toe." In *More Games of No Chance*, edited by R. J. Nowakowski, pp. 167–182, MSRI Publications 42. Cambridge, UK: Cambridge University Press, 2002.

[6] A. W. Hales and R. I. Jewett. "Regularity and Positional Games." *Transactions of the American Mathematical Society* 106:2 (1963), 222–229.

[7] O. Patashnik. "Qubic: 4 × 4 × 4 Tic-Tac-Toe." *Mathematics Magazine* 53:4 (1980), 202–216.

Scheduling Tennis Doubles Competitions

Dick Hess

Some time ago, a friend wrote to me that his club was organizing a mixed doubles competition involving 20 players (10 men and 10 women). Their plan was to play a series of rounds using five tennis courts. In each round, all the men and women are paired as partners, and two teams are assigned to each court to play each other in that round. It is required that no two players partner each other more than once or oppose each other more than once. My friend asked how many rounds could be played subject to these restrictions. He made it clear that it is OK for a player to partner someone in one round and oppose that same person in a different round. He had five rounds laid out as shown in Table 1, with men as odd numbers in the odd-numbered positions and women

Court 1	Court 2	Court 3	Court 4	Court 5
1 2 3 4	5 6 7 8	9 10 11 12	13 14 15 16	17 18 19 20
1 16 17 2	3 12 7 18	5 20 11 14	9 8 13 10	15 6 19 4
1 6 11 10	3 2 9 14	5 8 15 18	13 12 17 20	7 4 19 16
1 8 9 20	3 10 13 4	7 14 11 18	15 2 17 12	5 16 19 6
1 12 13 16	3 20 15 10	5 18 9 4	11 2 19 8	7 6 17 14

Table 1. A possible five-round mixed doubles schedule.

as even numbers in the even-numbered positions. He could do no better than five rounds.

A closely related problem applies to scheduling men's or women's doubles events in which there is no restriction that players partner with those of the opposite sex. The purpose of this paper is to report results for scheduling such tennis doubles events for 20 or fewer competitors and to compare results with those for the better known "social golf problem" with foursomes and the "bridge problem" for people playing duplicate bridge.

Related Problems

Scheduling problems are notoriously difficult to solve in the general case, so it is instructive to examine related problems. A very famous related problem is Kirkman's Schoolgirl Problem. initially involving threesomes to be scheduled, first published by Thomas Kirkman in 1847 and 1850 [5]. Soon the problem was expanded to foursomes and larger groups. Methods to solve the problem for various cases are given by Rouse Ball [2, pp. 267–298] and by Martin Gardner [3]. Today, the general problem is often referred to as the *social golf problem* and results are reported in [4, 6]. When stated with foursomes, the problem has $4m$ golfers split into m foursomes in each of several rounds. No two golfers may be in a foursome together more than once, and the problem is to determine the maximum number of rounds the golfers may play without violating the restriction. This problem is more restrictive than the tennis doubles problems because, in the doubles problems, two players may be on the same court with each other two times, once as partners and once as opponents. On the other hand, the mixed doubles problem has the added restriction that only a man and a woman are allowed to be partners.

A problem called the *bridge problem* in [2] asks that $4m$ members of a bridge club schedule $4m - 1$ rounds so that no two members partner each other more than once and each member opposes each other member exactly twice. This problem has been solved for many values of m, as reported in [2].

Upper Bounds

In the social golf problem, each round uses up three golfers playing with the first player. Thus, the number of rounds can never be more than $(4m - 1)/3$. For 8 or 12 golfers ($m = 2$ or 3), a further restriction applies: each of the four players of a foursome in the first round needs three new golfers to join in a possible second round. This need cannot be filled by only 4 or 8 remaining golfers, so only one round is possible with 8 or 12 golfers.

In the bridge problem, $4m - 1$ rounds is the upper bound and is achievable for many values of m.

In both types of doubles events with $4m$ players, each round uses up two players as opponents of the first player. Thus, the number of rounds can never be more than $2m - 1$.

Disk Method of Solution

Many approaches have been used to solve these scheduling problems. The one using disks, as discussed in [2] and [3], will be demonstrated here for eight players in the bridge and doubles problems. For the bridge problem, the approach is to represent seven of the players as points equally spaced on a circle, with the eighth player as a point in the center of the circle; see Figure 1. We join points 2 through 7 in pairs by solid-line chords to represent partnerships and by dashed-line chords to represent opponents. Points 1 and 8 are also joined by a solid line and by dashed lines to their opponents. If the solid chord lengths are all different and the dashed chord lengths are represented only twice each (as shown

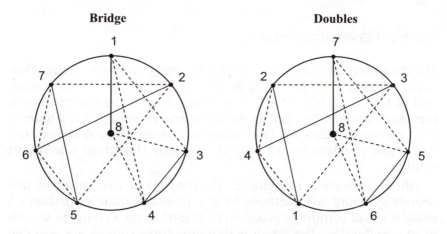

Figure 1. Disks for solving the eight-player bridge problem (left) and eight-player doubles problems (right).

Table 1	Table 2	Court 1	Court 2
8 1 3 4	2 6 5 7		
8 2 4 5	3 7 6 1	Court 1	Court 2
8 3 5 6	4 1 7 2	1 2 3 4	5 6 7 8
8 4 6 7	5 2 1 3	1 4 5 8	3 2 7 6
8 5 7 1	6 3 2 4	1 8 7 2	3 6 5 4
8 6 1 2	7 4 3 5		
8 7 2 3	1 5 4 6		

Table 2. The bridge and doubles problems for eight players.

in Figure 1), then a one-step cyclic permutation of the numbers 1 through 7 will give the assignments of the players for each of the seven rounds. The left half of Table 2 shows the assignments corresponding to the left half of Figure 1, giving the maximum number of rounds possible for the bridge problem.

The mixed doubles problem can use the same geometry as that for the bridge problem if the labels of the points are scrambled to keep partnerships only between even- and odd-numbered players, as shown in the right half of Figure 1. The maximum possible three rounds emerge by moving the numbers on the circle two spaces counterclockwise to generate Round 2 and then two more spaces to generate Round 3. The right half of Table 2 gives the result, which is also the best possible for men's or women's doubles.

Twelve-Player Problems

The disk method solves the 12-player bridge problem easily. One such solution from [2] is (12 & 1 vs. 5 & 6), (2 & 11 vs. 3 & 9), and (4 & 8 vs. 7 & 10) for the first round, with the ten remaining rounds developed by cycling the numbers 1 through 11. Figure 2 (left) shows the disk for this solution. As mentioned before, the social golf problem for 12 players has only the trivial one-round solution.

After attempting to achieve the theoretical maximum of five rounds by using disk methods for the 12-player doubles problem, I wrote a small computer program in BASIC to do a compete search of all possibilities. The result is that only three rounds are possible in mixed doubles and only four rounds are possible in men's or women's doubles. Table 3 shows solutions from the program.

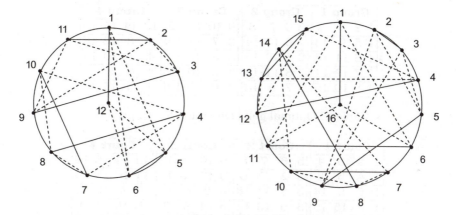

Figure 2. Disks for solving the 12-player and 16-player bridge problems.

Court 1	Court 2	Court 3
1 2 3 4	5 6 7 8	9 10 11 12
1 3 5 7	2 9 6 10	4 11 8 12
1 4 6 9	2 8 7 12	3 5 10 11
1 7 10 12	2 3 8 9	4 6 5 11

Court 1	Court 2	Court 3
1 2 3 4	5 6 7 8	9 10 11 12
1 4 5 8	3 6 9 12	7 2 11 10
1 6 11 2	3 12 5 4	7 10 9 8

Table 3. Optimal doubles scheduling for 12 players: men's or women's (top) and mixed (bottom).

Sixteen-Player Problems

The disk method solves the 16-player bridge problem; one such solution from [2] is (16 & 1 vs. 6 & 11), (2 & 3 vs. 5 & 9), (4 & 12 vs. 13 & 15), and (7 & 10 vs. 8 & 14) for the first round, with the 14 remaining rounds developed by cycling the numbers 1 through 15. The right half of Figure 2 shows the disk for this set of initial pairings. The theoretical optimum is achievable for the social golf problem with 16 players and is reported in [6]. Table 4 shows the result.

Remarkably, the theoretical maximum seven rounds can be reached for both types of doubles problems with 16 players. I could not find any solution on a disk (it may not be possible), but a computer program found the optimal solution given in Table 5. It has

Group 1	Group 2	Group 3	Group 4
1 2 3 4	5 6 7 8	9 10 11 12	13 14 15 16
1 5 9 13	2 6 10 14	3 7 11 15	4 8 12 16
1 6 11 16	2 5 12 15	3 8 9 14	4 7 10 13
1 7 12 14	2 8 11 13	3 5 10 16	4 6 9 15
1 8 10 15	2 7 9 16	3 6 12 13	4 5 11 14

Table 4. Optimal golf scheduling for 16 players.

Court 1	Court 2	Court 3	Court 4
1 2 3 4	5 6 7 8	9 10 11 12	13 14 15 16
1 4 5 8	3 2 7 6	9 12 13 16	11 10 15 14
1 6 9 14	3 8 11 16	5 2 13 10	7 4 15 12
1 8 15 10	3 6 13 12	5 14 9 2	7 16 11 4
1 10 13 6	3 12 15 8	5 4 11 14	7 2 9 16
1 12 11 2	3 10 9 4	5 16 15 6	7 14 13 8
1 14 7 12	3 16 5 10	9 6 15 4	11 8 13 2

Table 5. Optimal doubles scheduling for 16 players.

the interesting property that, if the players are eight married couples, no married couple ever needs to play together as a team. This property holds for any number of players if the theoretical maximum of $2m - 1$ rounds can be achieved. It is interesting that the theoretical optimum can be achieved for 16 players for each of the four problems studied here.

The 20-Player Problems

The disk method solves the 20-player bridge problem in many different ways; one such solution is (20 & 1 vs. 2 & 3), (4 & 6 vs. 10 & 16), (5 & 15 vs. 13 & 18), (7 & 19 vs. 11 & 14), and (8 & 12 vs. 9 & 17) for the first round, with the 18 remaining rounds developed by cycling the numbers 1 through 19. Figure 3 shows this solution on a disk. The theoretical optimum of six rounds for the social golf problem for 20 players cannot be achieved. Table 6 gives the best possible five rounds as produced from a complete search, as reported in [6].

The 20-player doubles problems have not been completely solved, to my knowledge. The disk method seems not to work, and complete searches take a long time to complete. My BASIC program that searches for solutions to each doubles scheduling problem ran for several weeks to produce the best known solutions in

A Lifetime of Puzzles

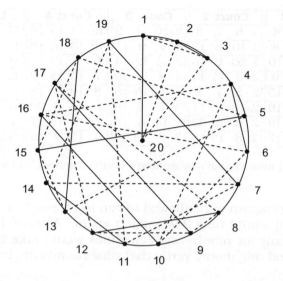

Figure 3. Disk for solving the 20-player bridge problem.

Group 1	Group 2	Group 3	Group 4	Group 5
1 2 3 4	5 6 7 8	9 10 11 12	13 14 15 16	17 18 19 20
1 5 9 13	2 10 15 17	3 8 14 20	4 7 12 19	6 11 16 18
1 7 11 15	2 9 16 20	3 6 13 19	4 8 10 18	5 12 14 17
1 8 12 16	2 11 14 19	3 5 15 18	4 6 9 17	7 10 13 20
1 6 10 14	2 12 13 18	3 7 16 17	4 5 11 20	8 9 15 19

Table 6. Optimal golf scheduling for 20 players.

Court 1	Court 2	Court 3	Court 4	Court5
1 2 3 4	5 6 7 8	9 10 11 12	13 14 15 16	17 18 19 20
1 4 5 8	3 2 7 6	9 12 13 16	11 10 17 20	15 14 19 18
1 6 9 14	3 8 11 16	5 2 13 10	7 20 19 4	15 18 17 12
1 8 7 18	3 10 15 4	5 16 11 20	9 2 17 14	13 12 19 6
1 16 17 10	3 14 13 20	5 18 9 6	7 4 15 12	11 8 19 2
1 18 13 2	3 6 19 10	5 4 17 16	7 12 11 14	9 8 15 20
1 20 15 6	3 12 17 8	5 10 19 14	7 2 9 16	11 4 13 18

Table 7. Best known mixed doubles scheduling for 20 players.

Tables 7 and 8. It shows a seven-round solution for mixed doubles and an eight-round solution for men's or women's doubles; neither is the theoretical maximum of nine rounds, so the question remains whether these solutions can be improved. I estimate that if

Court 1	Court 2	Court 3	Court 4	Court 5
1 2 3 4	5 6 7 8	9 10 11 12	13 14 15 16	17 18 19 20
1 3 5 8	2 10 6 17	4 7 9 11	12 13 14 18	15 19 16 20
1 4 6 10	2 20 8 16	3 11 12 14	5 7 13 17	9 15 18 19
1 5 2 9	3 6 11 13	4 8 14 19	7 10 15 20	12 18 16 17
1 6 12 15	2 3 7 9	4 5 16 18	8 19 10 13	11 14 17 20
1 7 14 16	2 18 10 11	3 5 4 19	6 13 9 20	8 15 12 17
1 11 13 19	2 5 12 20	3 14 6 18	4 9 15 17	7 16 8 10
1 16 11 17	2 8 15 18	3 9 10 20	4 12 7 13	5 19 6 14

Table 8. Best known men's or women's doubles scheduling for 20 players.

the BASIC program were allowed to run to completion on my home computer, it would take 244 million years. A restricted program searching only for nine-round schedules would take 214 years to complete and might only verify that nine rounds are impossible.

Conclusions

Scheduling foursomes to play golf, bridge, and mixed and men's or women's doubles events is a difficult problem, but it has been completely solved for between one and four foursomes. For five foursomes, the social golf and bridge problems are completely solved. For five-foursome mixed doubles, the best known solution has only seven rounds; for five-foursome men's or women's doubles, the best known solution has eight rounds. Schedules for more than 20 players have not been investigated.

Bibliography

[1] Thomas J. Kirkman. "Query 6." *Lady's and Gentleman's Diary* 48 (1850), 48.

[2] W. W. Rouse Ball. *Mathematical Recreations and Essays*, 13th edition. New York: Dover Publications, 1987. (First edition published in 1892.)

[3] Martin Gardner. "Mathematical Games." *Scientific American* 242 (May 1980), 16–28.

[4] Warwick Harvey. "Warwick's Results Page for the Social Golfer Problem." Available at http://www.icparc.ic.ac.uk/~wh/golf/, accessed 2006.

[5] Warwick Harvey. "The Fully Social Golfer Problem." In *Proceedings of the Third International Workshop on Symmetry in Constraint Satisfaction Problems* (SymCon'03), pp. 75–85, 2003.

When Multiplication Mixes Up Digits

David Wolfe

When I was in grade school, my father's workplace was about to throw out several decades of *Scientific American* magazines. He rescued them and delivered them to our basement. I would open each issue and flip to the "Mathematical Games" column, reading it with care. While school introduced me to arithmetic, Martin Gardner introduced me, along with countless other budding mathematicians, to mathematics.

Dr. Matrix was particularly fond of numbers consisting of all ten digits. In *The Magic Numbers of Dr. Matrix*, he suggests subtracting 123456789 from 987654321, and 0123456789 from 9876543210; both yield surprising answers. He also suggests puzzles such as inserting $+$ and $-$ signs in 123456789 to make 100. (See Solutions section at the end.)

Now that my daughter, Lila, is learning about counting, I recently wrote out the digits 1 through 9 on our whiteboard. She asked, "What number is that?" I explained, "Why, that's 123 million, 456 thousand, 789." She responded, "That's a very big number. Can you make a bigger one?" I doubled the number, and got

$$123456789 \times 2 = 246913578.$$

Wow! The product has the same digits 1 through 9, reordered. Before long, I found myself doubling it over and over again:

$$
\begin{array}{rcll}
123456789 & \times \ 2 \ = & 246913578, \\
246913578 & \times \ 2 \ = & 493827156, \\
493827156 & \times \ 2 \ = & 987654312, \\
987654312 & \times \ 2 \ = & 1975308624, \\
1975308624 & \times \ 2 \ = & 3950617248, \\
3950617248 & \times \ 2 \ = & 7901234496 & \text{(first exception).}
\end{array}
$$

Notice that every result is *pandigital*, until the last result, which has two 4s and two 9s. Why are there so many pandigital multiples?[1]

It turns out that the doubling process is a red herring, for lots of multiples of 123456789 are pandigital. If you list the numbers under 10 that, when multiplied by 123456789, are pandigital, you find

$$
\begin{array}{rcll}
123456789 & \times \ 1 \ = & 123456789, \\
123456789 & \times \ 2 \ = & 246913578, \\
123456789 & \times \ 4 \ = & 493827156, \\
123456789 & \times \ 5 \ = & 617283945, \\
123456789 & \times \ 7 \ = & 864197523, \\
123456789 & \times \ 8 \ = & 987654312,
\end{array}
$$

that is, all the single-digit numbers that don't have a prime factor of 3.

To investigate what exactly is going on, we generalize the question to base b:

Theorem 23.1. *Let x be the base-b number $123\ldots(b-1)$, and choose an n between 1 and b. The product $n \cdot x$ is pandigital if and only if $b-1$ and n share no prime factors.*

In particular, in base $b = 10$, we have $b-1 = 9$, which has a single prime factor, 3. The theorem says that for values of n that *don't* have a factor of 3, i.e., when n is 2, 4, 5, 7, or 8, multiplication by n results in a pandigital product.

Using a diagram, we can compute the product another way by "walking around a clock," and in so doing can shed light on the theorem. After describing this clock method, we'll see why it explains the theorem and then why the clock method correctly computes the product.

---

[1]A pandigital number in base b contains all the base-b digits. The literature varies about whether 0 needs to be one of the digits. Here we say a number is pandigital if it contains either all the digits *or* all the nonzero digits.

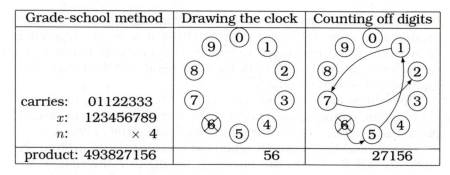

Grade-school method	Drawing the clock	Counting off digits
carries: 01122333 x: 123456789 n: × 4		
product: 493827156	56	27156

Figure 1. The clock method.

To explain the clock method, we'll walk through an example of multiplying 123456789 by $n = 4$. On the left in Figure 1 is the usual method for multiplying that we learned in grade school.

For the clock method, first write the digits 0 through $b - 1$ in a circle. The last two digits in the product are $b - n$ (in the example, $b - n = 6$) and $b - n - 1$ (which is 5). Write these down in the product. Cross out the $b - n = 6$. Now, beginning with $b - n - 1 = 5$, count counterclockwise $n = 4$ positions around the circle to read off the digits in the product. In the example, you'll go from 5 to 1 to 7 to 2 to 8 to 3 to 9 to 4 to 0. Don't count the crossed-out $b - n = 6$. Upon recording the nine-digit product, stop; the tenth digit will be 0.

This process hits all the nonzero digits if and only if $b - 1$ (the number of digits not crossed out) and n share no common factors. (Otherwise, you'd repeat digits as you used the clock method.)

To see why this alternative way of computing the product works, let us compare what happens when you multiply 123456789 by small numbers like 4, using both the clock method and the grade-school method. First, look at the carries. The carry from the last digit is $n - 1 = 3$, and the carries stay the same or decrease, proceeding leftward from digit to digit. In our example on the left, $123456789 \cdot 4$, the digit products with carries are, in order *from right to left*,

$$36, \quad 35, \quad 31, \quad 27, \quad 22, \quad 18, \quad 13, \quad 9, \quad 4.$$

Suppose that while computing the product by the grade-school method, we forgot to carry. Because, working from right to left, the digits of 123456789 decrease by 1, each digit's product by $n = 4$ would decrease by $n = 4$. Reintroducing the carries, because the rightmost digit has no carry but generates a carry of $n - 1 = 3$, the last two digits of the product will differ by $n - (n - 1) = 1$.

Doing arithmetic modulo b, proceeding from right to left, the digit products decrease by 1 and then by either n or $n + 1$, depending on whether the carry stayed the same or decreased. Further, the carry decreases when the ten's (or, in general, b's) digit decreases and the unit's digit increases.

This brings us to why we cross out the 6. Returning to the circle of digits, counting by 4 reflects the fact that consecutive digit products differ by 4 or 5. They differ by 5 when the previous carry decreased, and that's exactly when the previous count around the circle passed 0. If you are currently on digits 7 through 9, the carry must have just dropped, and the next product should decrease by 5 rather than 4. Crossing out the 6 is tantamount to counting down 5 rather than 4 when the current digit is 7 through 9, because the next count will skip the 6.

We can use the same method of analysis for 987654321 and its generalization to base b:

Theorem 23.2. *Let x be the base-b number $(b-1)\ldots 321$ and choose an n between 1 and b. The product $n \cdot x$ is pandigital if and only if $b - 1$ and n share no prime factors.*

Here we use a slightly different clock process to generate the product. Start with the same circle of digits. Cross out n, and write it down as the rightmost digit. Then, count by ns (skipping the crossed-out n) *clockwise* until all b digits are recorded.

Returning to Theorem 23.1, 123456789 times n is pandigital for lots of larger values of n, too. In particular, n can be any of 10, 11, 13, 14, 16, 17, 20, 22, 23, 25, 26, 31, 32, 34, 35, 40, 41, 43, 44, 50, 52, 53, 61, 62, 70, 71, or 80 (and no other two-digit number). For instance,

$$123456789 \times 71 = 8765432019.$$

Note that this list includes no numbers that are a multiple of 3 (which comes as no surprise) but also omits other numbers, such as 19. We leave it open to generalize this example to base b.

Note: The fact that multiples of 123456789 and 987654321 are pandigital has long been observed. See, for example, David Wells' *The Penguin Dictionary of Curious and Interesting Numbers* (Penquin, 1986), or do a web search on "123456789 pandigital." The author is surprised not to have seen the generalization to base b.

Acknowledgments. Thanks to David Molnar, who identified that the multipliers yielding pandigital numbers are relatively prime to $b - 1$ in Theorem 23.1.

These results were first published in *Mathematics Magazine*, December 2007, and are reprinted here with permission.

Bibliography

[1] Martin Gardner. *The Magic Numbers of Dr. Matrix.* New York: Prometheus Books, 1985.

Solutions

The first difference is

$$987654321 - 123456789 = 864197532,$$

which is pandigital.

The second difference is

$$9876543210 - 0123456789 = 9753086421,$$

which is also pandigital.

There are 11 ways to insert + and − signs in 123456789 to make 100:

1+2+3−4+5+6+78+9, 1+2+34−5+67−8+9, 1+23−4+5+6+78−9,
1+23−4+56+7+8+9, 12+3+4+5−6−7+89, 12+3−4+5+67+8+9,
12−3−4+5−6+7+89, 123+4−5+67−89, 123+45−67+8−9,
123−4−5−6−7+8−9, 123−45−67+89.

Magic, Antimagic, and Talisman Squares

Rodolfo Kurchan

Magic squares are a classic topic in recreational mathematics. Martin Gardner discusses them in 12 of his 15 books on mathematical games; see, for example, his article "Magic Squares and Cubes" [1]. Here I describe some modern variations on magic squares and my findings about these squares by myself and others. All of my solutions were found without using computers.

Pandigital Magic Squares

A *magic square* is an $n \times n$ array of numbers such that the rows, columns, and main diagonals produce the same sum, called the *magic sum*. Here is a simple example, with a magic sum of 15:

6	1	8
7	5	3
2	9	4

A decimal number is *pandigital* if it uses precisely all ten digits and zero is not the leading digit. A *pandigital magic square* consists only of pandigital numbers and has a pandigital magic sum. In 1989, Rudolf Ondrejka [5] posed this problem: what is the pandigital magic square with the smallest pandigital magic sum?

In 1991, I found this solution [2], with pandigital magic sum 4,129,607,358:

1,037,956,284	1,036,947,285	1,027,856,394	1,026,847,395
1,026,857,394	1,027,846,395	1,036,957,284	1,037,946,285
1,036,847,295	1,037,856,294	1,026,947,385	1,027,956,384
1,027,946,385	1,026,957,384	1,037,846,295	1,036,857,294

In 2003, I found an improved solution [3], with pandigital magic sum 4,120,736,958:

1,034,728,695	1,035,628,794	1,024,739,685	1,025,639,784
1,024,639,785	1,025,739,684	1,034,628,795	1,035,728,694
1,035,629,784	1,034,729,685	1,025,638,794	1,024,738,695
1,025,738,694	1,024,638,795	1,035,729,684	1,034,629,785

In 2004, Carlos Rivera [6] quested for the smallest 3×3 pandigital magic square. By a computer search, he found the following great minimal solution, with pandigital magic sum 3,205,647,819:

1,057,834,962	1,084,263,579	1,063,549,278
1,074,263,589	1,068,549,273	1,062,834,957
1,073,549,268	1,052,834,967	1,079,263,584

Some interesting open problems about pandigital magic squares are the following:

(a) What is the pandigital magic square with the largest sum?

(b) Is there a 5×5 or larger pandigital magic square?

Antimagic Squares

An *antimagic square* is an $n \times n$ array of the numbers from 1 to n^2 such that the rows, columns, and main diagonals produce different sums, and the sums form a consecutive sequence of integers. Antimagic squares were invented by J. A. Lindon in 1962 [4, p. 103]. Here is an example:

				29
6	8	9	7	**30**
3	12	5	11	**31**
10	1	14	13	**38**
16	15	4	2	**37**
35	**36**	**32**	**33**	**34**

A Lifetime of Puzzles

In 2004, I found this 5×5 antimagic square that contains in its center a 3×3 magic square [7]:

					59
7	8	24	22	2	**63**
4	16	9	14	21	**64**
25	11	13	15	5	**69**
6	12	17	10	23	**68**
18	20	3	1	19	**61**
60	**67**	**66**	**62**	**70**	**65**

In 2005, I found a 6×6 antimagic square that contains in its center a 4×4 magic square:

						108
1	36	34	33	2	3	**109**
35	26	13	12	23	6	**115**
27	15	20	21	18	5	**106**
10	19	16	17	22	30	**114**
9	14	25	24	11	29	**112**
31	7	8	4	28	32	**110**
113	**117**	**116**	**111**	**104**	**105**	**107**

An interesting open problem is to find an $n \times n$ magic square that contains in its center an $(n-2) \times (n-2)$ antimagic square, or to prove that this is impossible.

Talisman Squares

The *Talisman constant* of an $n \times n$ array of the numbers from 1 to n^2 is the minimum difference between any element and one of its eight immediate neighbors (including diagonal neighbors). A *Talisman square* is an $n \times n$ array with the largest possible Talisman constant over all $n \times n$ arrays of the numbers from 1 to n^2. Talisman squares were invented by Sidney Kravitz [4, p. 110]. As an example, consider the following two 4×4 squares:

16	3	2	13
5	10	11	8
9	6	7	12
4	15	14	1

9	5	11	7
13	1	15	3
10	6	12	8
14	2	16	4

The left square is the well-known Dürer Magic Square (appearing in Dürer's *Melancholia I* engraving from 1514) and has a Talisman constant of 1. On the other hand, the right square has a

Talisman constant of 3. So, the left square cannot be a Talisman square, while we may assert that the right square is a Talisman square because 3 is the largest Talisman constant possible for any 4×4 square.

But how do we construct an $n \times n$ Talisman square for any given order n? Carlos Rivera and I [8] have been studying this problem, and we have found a pair of algorithms (one algorithm for even values of n, the other for odd n) that we conjecture produce a Talisman square for any given order n, as desired. Instead of providing long-winded and boring general instruction rules, we will demonstrate the algorithms using a couple of examples and some explanations about them.

Even n

We obtain a Talisman constant of $n^2/4 - 1$ for even n. Here is an example of our algorithm applied to $n = 6$, where the Talisman constant is 8:

19	10	22	13	25	16
28	1	31	4	34	7
20	11	23	14	26	17
29	2	32	5	35	8
21	12	24	15	27	18
30	3	33	6	36	9

As you may have noticed, the filling pattern in the previous example divides the numbers $1, 2, 3, ..., n^2$ into four sets (S_1, S_2, S_3, S_4), each with $n^2/4$ consecutive numbers, as follows:

$$
\begin{aligned}
S_1 &= \{1, & 2, & 3, & ..., & X - 1\}, \\
S_2 &= \{X, & X + 1, & X + 2, & ..., & Y - 1\}, \\
S_3 &= \{Y, & Y + 1, & Y + 2, & ..., & Z - 1\}, \\
S_4 &= \{Z, & Z + 1, & Z + 2, & ..., & n^2\}.
\end{aligned}
$$

The $n^2/4$ consecutive numbers of each set are allocated in the same general trend:

> Starting from a certain specific position inside the four cells in the upper-left corner, the rest of the numbers of each set are allocated consecutively according to the rule "every two cells downward and rightward." At the moment you finish allocating the last number of the first set, you know the first number of the following set, and so on.

So, the only important thing you should know in advance is the cells in which the first numbers of each set $(1, X, Y, \text{and } Z)$ must be allocated. The answer is this:

1 goes in the cell $(2, 2)$,

X goes in the cell $(1, 2)$,

Y goes in the cell $(1, 1)$, and

Z goes in the cell $(2, 1)$,

as shown below. Moreover, if you want to know the values in advance, $X = n^2/4 + 1$, $Y = n^2/2 + 1$, and $Z = 3n^2/4 + 1$, but this is not really necessary to know.

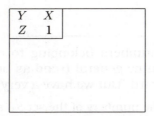

We call this filling pattern "22A": "22" because it starts in the cell $(2, 2)$, and "A" because the four starting numbers of each set— 1, X, Y, and Z—describe the profile of a letter "A."

Odd *n*

We obtain a Talisman constant of $\lfloor n(n-1)/4 \rfloor$, the integer immediately below $n(n-1)/4$, for odd n. Here is an example of our algorithm applied to $n = 7$, where the Talisman constant is 10:

13	40	17	32	21	36	25
1	29	4	44	7	47	10
14	41	18	33	22	37	26
2	30	5	45	8	48	11
15	42	19	34	23	38	27
3	31	6	46	9	49	12
16	43	20	35	24	39	28

As before, there are four sets (S_1, S_2, S_3, S_4) of consecutive numbers. Now, however, the four sets have distinct quantities of integers. Again, the starting numbers of the four sets, 1, X, Y, and Z, are allocated in the four cells in the upper-left corner, but now

$$1 \text{ goes in the cell } (2, 1),$$
$$X \text{ goes in the cell } (1, 1),$$
$$Y \text{ goes in the cell } (2, 2), \text{ and}$$
$$Z \text{ goes in the cell } (1, 2),$$

as shown below. We call this pattern "21N" for analogous reasons as before.

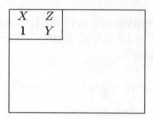

The consecutive numbers belonging to each of the four sets are allocated in the same general trend as before: every two cells, downward and rightward. But we have a very important difference:

> When allocating the numbers of the set S_3, starting in column $4 + 2(\lfloor c/4 \rfloor - 1)$, shift upward by one cell all the cells that would receive the corresponding numbers for this column. The same happens with all columns rightward of this column.

> Consequently, when allocating the numbers of the set S_4, starting at column $4 + 2(\lfloor c/4 \rfloor - 1)$, shift downward by one cell all the cells that would receive the corresponding numbers for this column. The same happens with all columns rightward of this column.

Summary

Talisman squares are constructed as follows.

- For n even:

    Use the filling pattern 22A[1] for the starting numbers $(1, X, Y, \text{ and } Z)$ of the four sets $(S_1, S_2, S_3, \text{ and } S_4)$ of $n^2/4$ consecutive numbers; allocate the numbers of each set using the general procedure "every two cells downward, rightward." Proceeding this way, we obtain a Talisman constant of $n^2/4 - 1$.

---

[1] In fact, for even n, we have found two more general patterns that produce the same Talisman constant. We have selected the pattern 22A because it seems appropriate for producing Talisman rectangles, as well. However, this is still a work in progress.

- For n odd:

  Use the filling pattern 21N for the starting numbers $(1, X, Y,$ and $Z)$ of the four sets $(S_1, S_2, S_3,$ and $S_4)$ of consecutive numbers; allocate the numbers of each set using the general procedure "every two cells downward, rightward." For the sets S_3 and S_4, shift upward and downward, respectively, the starting cell in each column from $4 + 2(\lfloor c/4 \rfloor - 1)$ rightward. Proceeding this way, we obtain a Talisman constant of $\lfloor n(n-1)/4 \rfloor$.

n, order of Talisman square	3	4	5	6	7	8	9	10	11
Talisman constant:									
$n^2/4 - 1$, for n even;		3		8		15		24	
$\lfloor n(n-1)/4 \rfloor$ for n odd.	1		5		10		18		27
First shifted column:									
$4 + 2(\lfloor c/4 \rfloor - 1)$,	–	–	4	–	4	–	6	–	6
sets S_3 and S_4, just for n odd.									

It is an open problem whether our Talisman constants can be improved, or whether our constructions are indeed Talisman squares. In May 2004, Luke Pebody [8] proved that our algorithm produces Talisman squares for even n. But, the situation seems significantly more complicated for odd n.

Bibliography

[1] Martin Gardner. *Time Travel and Other Mathematical Bewilderments*. New York: W. H. Freeman, 1988.

[2] Rodolfo Kurchan. "Solution to Problem 1694." *Journal of Recreational Mathematics* 23:1 (1991), 69.

[3] Rodolfo Kurchan's home page. Available at http://www.snarkianos.com/rodolfo/rodolfo.html, 2003.

[4] Joseph S. Madachy. *Madachy's Mathematical Recreations*. New York: Dover Publications Inc., 1979. Originally published as *Mathematics on Vacation*. New York: Charles Scribner's Sons, 1966.

[5] Rudolf Ondrejka. "Problem 1694." *Journal of Recreational Mathematics* 21:1 (1989), 68.

[6] Carlos Rivera. "Puzzle 249." *The Prime Puzzles & Problems Connection*. Available at http://www.primepuzzles.net/puzzles/puzz_249.htm, 2004.

[7] Carlos Rivera. "Puzzle 263." *The Prime Puzzles & Problems Connection.* Available at http://www.primepuzzles.net/puzzles/puzz_263.htm, 2004.

[8] Carlos Rivera. "Puzzle 267." *The Prime Puzzles & Problems Connection.* Available at http://www.primepuzzles.net/puzzles/puzz_267.htm, 2004.

Rectangle Arithmetic: Another Slant on Fractions

Bill Gosper

This article describes a geometric view of numbers that connects grade-school arithmetic, geometric slopes, continued fractions, and electrical resistance.

For starters, boxes represent numbers. A square, ■, regardless of size, has value 1:

For the original, color version of this article, please see http://www. tweedledum.com/rwg/rectarith12.pdf. It is much clearer and prettier and may yet appear in print.

A stack of two squares, 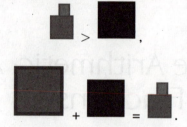, means 2, again regardless of their sizes:

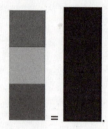

A stack of equal squares can form a rectangle:

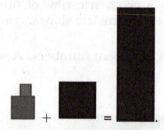

This means $1 + 1 + 1 = 3$. So,

Size doesn't matter—only shape. We add rectangles by stacking them vertically, like with squares:

Here, two equal rectangles sum to 1. So, if the rectangles each represent x, we have $x + x = 1$. Thus, $x = \frac{1}{2}$, which we can see as 2 turned sideways.

A Lifetime of Puzzles

A tall rectangle that is not a whole number of squares is an "improper" fraction. By marking off the squares, we make a mixed number:

In symbols, $\frac{5}{2} = 2 + \frac{1}{2}$.

A 3 turned sideways is obviously $\frac{1}{3}$. And $\frac{2}{3}$ is

Turning this sideways, we get $\frac{3}{2}$:

Turning any rectangle sideways reciprocates its value. In general, the value represented by a rectangle is just its height divided by its width, that is, the slope of its diagonal. Engineers use the fancy term "aspect ratio," but they like to make it greater than or equal to one by sometimes switching height and width. But, if we did that, we'd confuse 3 with $\frac{1}{3}$!

So, rectangles just represent fractions: the height is the numerator, and the width is the denominator. The reason size doesn't matter is that magnifying a rectangle is the same as multiplying the numerator and denominator by the same quantity.

When adding $\frac{1}{3} + \frac{1}{2}$, scaling the summands to have the same width and making a nice rectangle is the same as finding a common denominator: for example,

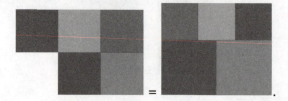

This equation, and the whole idea that shape matters, but not size, may seem artificial and slapdash, but there is actually a simple physical example of this behavior. If each square is made of the same electrically resistive material, and we coat their top and bottom edges with a good conductor and then apply a voltage between the topmost and bottommost edges of the above figures, the currents that they pass will be equal and will not change if the figures are scaled up or down. To help your intuition, suppose that you have a square conducting a certain current. Placing another square beside it (creating the rectangle value $\frac{1}{2}$) doubles the current, and thus halves the resistance:

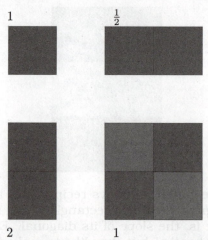

But then stacking two of these rectangles vertically creates a large square and redoubles the resistance back to the initial value.

A Lifetime of Puzzles

We can subtract a smaller rectangle from a larger one by scaling to equal width (finding a common denominator) and lopping off the smaller from the larger:

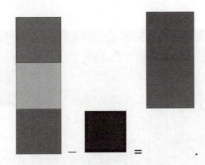

But the extra machinery that we'd need for handling negative numbers probably isn't worth it. We could draw a diagonal along each rectangle. The opposite diagonal means the opposite sign; vertical flipping reverses the sign. To add values, scale them to equal width, and never let diagonals join end to end; instead, superpose the rectangles to have a common upper or lower edge. Then, the sum is the rectangle whose diagonal joins the beginning of one diagonal with the end of the other. We'll avoid this complication in the rest of this article by focusing on positive numbers.

From the $\frac{1}{2} + \frac{1}{3}$ example, we can read off the answer $\frac{5}{6}$, i.e., 5 high and 6 wide, if we subdivide one of the larger squares (say, the lower left) by trisecting its edges:

A more methodical way to read off a rectangle's value is to convert it to a mixed number, reciprocate the fraction, and repeat:

We removed zero squares in the vertical dimension, because the fraction was "proper." Then, we got one square horizontally. Then, we got five squares vertically with no remainder, so the process (known as a continued fraction) terminated with the value

$$0 + \cfrac{1}{1 + \cfrac{1}{5}} = \frac{5}{6}.$$

When the slope is not a rational fraction, the continued fraction process does not terminate, as with the number π:

$$\pi = 3 + \cfrac{1}{7 + \cfrac{1}{15 + \cfrac{1}{1 + \cfrac{1}{292 + \cfrac{1}{\ddots}}}}},$$

Due to limited resolution, the column of 15 (more nearly 16) is barely visible. We call the sides of such a rectangle *incommensurable* because there is no scale of measurement in which both are whole numbers. Scaling a rectangle doesn't affect the commensurability of its sides.

What does it mean to stack rectangles horizontally rather than vertically? In other words, what is the slope of a rectangle joined by stacking equally tall rectangles horizontally? The answer is easy: if we turn it sideways, it's the sum of the reciprocals. Reciprocating the sum of the reciprocals (*harmonic sum*) is also how you add resistors in parallel. If we add two frequencies or angular velocities, we harmonically sum the periods (and wavelengths). For example, the time it takes the stars to fully circle Polaris (The North Star) is

$$1 \text{ sidereal day} = \cfrac{1}{\cfrac{1}{1 \text{ year}} + \cfrac{1}{1 \text{ day}}}$$

because the Earth both rotates and revolves.

Here is an algebra exercise. The following rectangle is composed of nine unequal squares. Is it a perfect square?

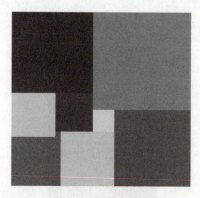

If not 1, what number does the rectangle represent? (Hint: Arbitrarily assign the value 1 to the sides of the tiniest square, and the value x to the square just above it. Then, their left neighbor has side $x + 1$. You can continue assigning sizes in terms of x to all nine squares. Then, equate the top edge of the rectangle with the bottom, and you should get an equation that determines x.) Answer: $x = 7$, and the rectangle has slope $\frac{32}{33}$.

For a dozen or so more of these diagrams, see [2]. For hundreds more (with the sizes filled in), search the web for "squared rectangles." If I were king, one of these diagrams (undimensioned) would appear daily in the newspaper puzzle pages, along with the answer to "Yesterday's Answer." Sundays would feature an extra-large one requiring simultaneous equations. Related is the ancient problem of finding a squared (or "perfect") *square*, covered by Martin Gardner [1] and first solved in the late 1930s.

Besides adding and subtracting, it's easy to *multiply* slopes. If four rectangles fit to form a larger one,

then the products of the diagonally opposite slopes are equal. In other words, upper left × lower right = lower left × upper right. This gives the term *cross product* a whole old meaning. In the example

A Lifetime of Puzzles

above, the upper left is square, and we get that $2 \times \frac{1}{3} = \frac{2}{3}$ (lower left × upper right = lower right). Making the lower-left square instead, we have $\frac{1}{3} \times \frac{1}{2} = \frac{1}{6}$ (lower right × upper left = upper right):

For *division*, just put the dividend diagonally opposite the 1 square, or reciprocate the multiplier.

We can represent the number $0 = \frac{0}{1} = \frac{0}{2} = \ldots$ with a horizontal stroke,—, i.e., a box of zero height. You can safely reciprocate it to make $\frac{1}{0} = \frac{2}{0} = \ldots$, represented by a harmless vertical stroke, |. We see that stacking up zeroes makes no difference, and multiplying by | makes |, except that multiplying | by — makes a single point, $\cdot = \frac{0}{0}$, and trying to do anything with this just makes another ·.

Here is how to take the *square root* of a slope. Let B stand for Bond. James Bond. He and Auric Goldfinger, G, creep out from opposite corners of the rectangular quarters of M. Slope M. They creep at identical speeds, so that $JB = AG$. As soon as they can see each other along the diagonal of a rectangle, they shoot. The line of fire JA has slope $\sqrt{M}$.

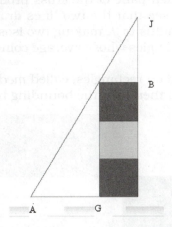

In this illustration, $M = 3$, creating half an equilateral triangle. But how do we actually synchronize Bond's motion with Goldfinger's to find the line JA? Without scaling, adjoin to the rectangle a sideways copy, as shown below.

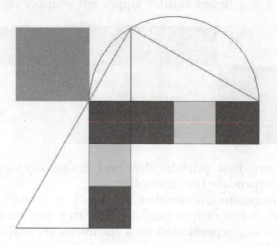

Join the two upper corners of the combined figure with a semi-circle. This intersects Bond's vertical path at the desired point J, from which we draw straight lines through those upper corners. The lines will be perpendicular, with the desired (reciprocal) slopes $s_1 = \sqrt{M}$ and $s_2 = -\frac{1}{\sqrt{M}}$, as shown by the top-left square (slope 1) and the multiplication rule: $1 \times M = s_1 \times s_1$. In fact, we don't even need the top-left square to see this: instead, use the elbow square as the lower-left pane of the cross product $1 \times \frac{1}{s_1} = s_1 \times \frac{1}{M}$ (ignoring signs). (To see that the two lines drawn from J are perpendicular, draw a radius to J, making two isosceles triangles with supplementary apex angles whose average coincides with the angle at J.)

One last operation on rectangles, called *mediant*, is to join them corner-to-corner and then take the bounding box:

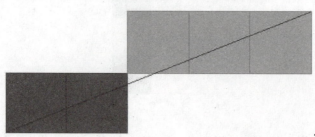

The above example says mediant$(\frac{1}{2}, \frac{1}{3}) = \frac{2}{5}$. This is the fraction

A Lifetime of Puzzles

between $\frac{1}{2}$ and $\frac{1}{3}$ that has the smallest numerator and denominator. In general, the mediant is the sum of the numerators over the sum of the denominators—the way you're *not* supposed to add fractions.

You can find the best rational approximations to any number between zero and infinity by repeatedly taking the mediant of the last underestimate and overestimate, starting with an underestimate — (zero) and an overestimate | (infinity). For example, what's the easiest way to bat 0.239? Write the underestimates on the left and the overestimates on the right, working toward the middle:

$$\frac{0}{1} \qquad\qquad\qquad\qquad\qquad\qquad\qquad\qquad \frac{1}{0}$$

$$\frac{1}{5} \qquad\qquad\qquad\qquad\qquad\qquad \frac{1}{4}\ \ \frac{1}{3}\ \ \frac{1}{2}\ \ \frac{1}{1}$$

$$\frac{2}{9}\ \ \frac{3}{13}\ \ \frac{4}{17}\ \ \frac{5}{21} \qquad\qquad \frac{6}{25}$$

$$\frac{11}{46}$$

(In decimals, $\frac{5}{21} \approx 0.2381$, and $\frac{11}{46} \approx 0.2391$.) So, you need (at least) 46 at-bats to bat 0.239. Graphically, a slope of exactly 0.239 requires a rectangle with one corner at $(0,0)$ and the opposite at $(1000, 239)$, way off the page. We chain together a | ($\frac{1}{0}$), putting us too high. Then, five —'s ($\frac{0}{1}$) put us at $(5, 1)$, too low. Then, five $\frac{1}{4}$'s put us a bit too high, and finally, a $\frac{5}{21}$ gets us within 0.0005:

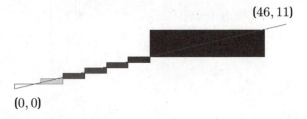

$(46, 11)$

$(0, 0)$

The big rectangle in the top right is the same as the box bounding $(0,0)$ and the first four $\frac{1}{4}$ rectangles. If we continue the process, we will soon reach the point $(1000, 239)$, which is the first time the slope will be exactly right, because $\frac{239}{1000}$ is in lowest terms. The grid point nearest $(0,0)$ through which a diagonal passes is its slope in lowest terms. This mediant process finds that point, thus reducing fractions without first finding the greatest common divisor. For example, for $\frac{65}{286} \approx 0.2273$,

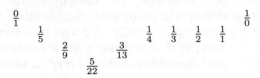

$$\frac{0}{1} \qquad\qquad\qquad\qquad\qquad\qquad\qquad\qquad \frac{1}{0}$$

$$\frac{1}{5} \qquad\qquad\qquad\qquad \frac{1}{4}\ \ \frac{1}{3}\ \ \frac{1}{2}\ \ \frac{1}{1}$$

$$\frac{2}{9}\ \ \frac{3}{13}$$

$$\frac{5}{22}$$

where the $\frac{5}{22}$ is in the middle because it's exact. The "usual" way to do this graphically is to repeatedly eat off the largest square from a 65-by-286 rectangle, which will terminate with the removal of a 13-by-13 square, which is the greatest common divisor, which you then divide out:

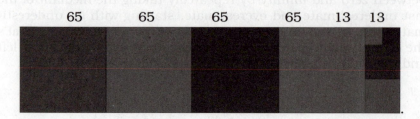

65	65	65	65	13	13

Instead of rectangles, it is also possible to compute mediants with circles. Below is a perspective view of "Euclid's orchard" as described in the "Lattice of Integers" chapter of *Martin Gardner's Sixth Book of Mathematical Diversions from Scientific American* (University of Chicago Press, 1984).

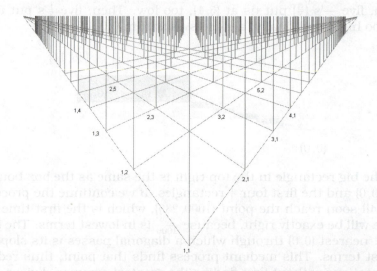

Notice how only grid points with coprime coordinates are "visible." Grid points such as $(2, 2)$ or $(2, 4)$ are hidden behind trees growing out of the corresponding grid point with the common divisor scaled out. If we flip this drawing vertically, it turns into the graph of the notorious "ruler" function that manages to be continuous at every irrational number and discontinuous at every rational one.

A Lifetime of Puzzles

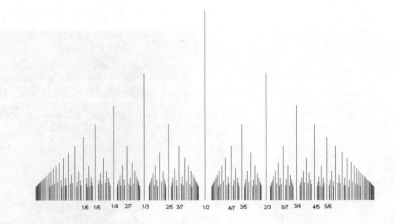

The exact definition is simply 0 at every irrational and $\frac{1}{d}$ for every rational $\frac{n}{d}$. Now adjoin points 0 and 1, where the function is 1, and square it.

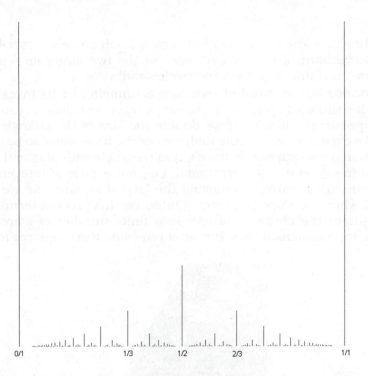

Finally, let each of these vertical segments be the diameter of a circle.

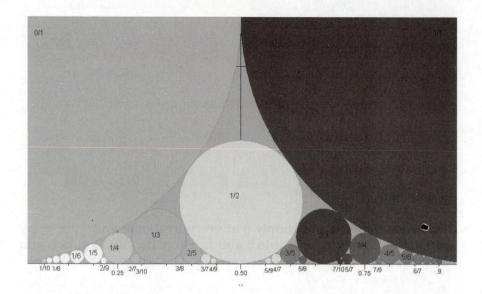

These are known as the *Ford circles*. Each circle's x-coordinate is the mediant of the x-coordinates of the two larger circles that confine it. If this is $\frac{n}{d}$, then the circle's radius is $\frac{1}{d^2}$.

For our last example of rectangle arithmetic, let us investigate whether three grid points (i.e., having integer coordinates) can form an equilateral triangle. If we double the size of the triangle (and maybe even if we don't), the midpoint of the base will also be a grid point, and we supposedly have a grid rectangle with diagonal slope equal to $\sqrt{3}$. But, $\sqrt{3}$ is irrational, i.e., not a ratio of integers. To see this, try iteratively removing the largest square. As we have seen, when the slope is a ratio of integers, this process terminates (consumes the entire rectangle) in a finite number of steps; the sides are commensurable. But after removing three squares for $\sqrt{3}$,

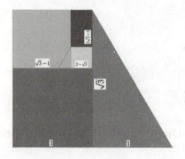

A Lifetime of Puzzles

we have a rectangle (in the top middle) with diagonal slope

$$\frac{2\sqrt{3} - 3}{2 - \sqrt{3}} = \frac{(2 - \sqrt{3})\sqrt{3}}{2 - \sqrt{3}} = \sqrt{3}.$$

In other words, this rectangle has the same diagonal as the original, so the process will go on forever. In fact, this gives the infinite continued fraction for $\sqrt{3}$:

$$\sqrt{3} = 1 + \cfrac{1}{1 + \cfrac{1}{1 + \sqrt{3}}} = 1 + \cfrac{1}{1 + \cfrac{1}{2 + \cfrac{1}{1 + \cfrac{1}{2 + \cfrac{1}{1 + \cfrac{1}{\ddots}}}}}}.$$

So, in the grid there is no equilateral triangle with a horizontal base. But what about tilting the triangle at some angle? Surprise! The slopes of *all* of the angles in the infinite, two-dimensional grid are rational and are thus found among the (untilted) grid rectangles. In fact, if two lines through a point have slopes s and t, the slope of the angle between them is

$$\frac{s - t}{1 + st},$$

which is clearly rational when s and t are both rational. But, instead of deriving this formula, there is a more intuitive way to see that there are no new angles to be had by tilting. As an example, we'll use the angle between slopes 3 (upper diagonal line) and $\frac{1}{5}$ (lower diagonal line), but the argument works in general.

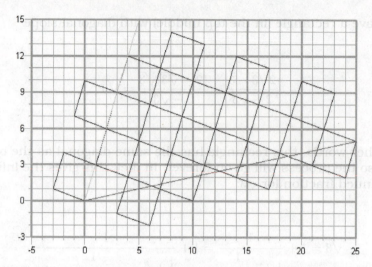

Arbitrarily choosing the slope-3 line, make a square grid using increments (sides of the squares) $(1, 3)$ and $(3, -1)$. Now follow the slope-$\frac{1}{5}$ line from $(0, 0)$. The heavier vertical grid lines interrupt it at $(5, 1)$, $(10, 2)$, $(15, 3)$, and so on. Notice that $(0, 0)$ is the corner of a square-grid square, and $(5, 1)$, $(10, 2)$, and so on land in various places inside these squares. But there are at most $10 = 3 \times 3 + 1 \times 1$ (in this case) different places to land before winding up on another square corner, in this case, $(25, 5)$. But then $(0, 0)$, $(4, 12)$, and $(25, 5)$ form half a rectangle (with a slope-$\frac{1}{5}$ diagonal) in the square grid, and by counting these squares, the angle between the slope-3 and slope-$\frac{1}{5}$ lines has slope $\frac{7}{4}$, just as predicted by the formula:

$$\frac{3 - \frac{1}{5}}{1 + 3(\frac{1}{5})} = \frac{14}{8} = \frac{7}{4}.$$

Notice that the sequence $(0, 0)$, $(5, 1)$, $(10, 2)$, ..., $(25, 5)$ visits exactly half of the ten possible grid points in the square grid. If we simulate wraparound by subtracting multiples of the edges $(1, 3)$ and $(3, -1)$ so as to confine the moving point to one square, we have the sequence

$$(0, 0), (1, -1), (2, -2), (3, -3) = (0, -2), (1, -3), (2, -4) = (-1, -3) = (0, 0).$$

Thus, for any pair of grid points, there is a rectangle of grid points with one vertex at the origin, one side through the first point, and one of its diagonals through the second point.

Let us finally escape from Flatland and idly ask, "Are there equilateral triangles in the *three*-dimensional grid?" Abundantly! With coordinates limited to just 0, 1, and 2, there are 24, as well as four regular hexagons.

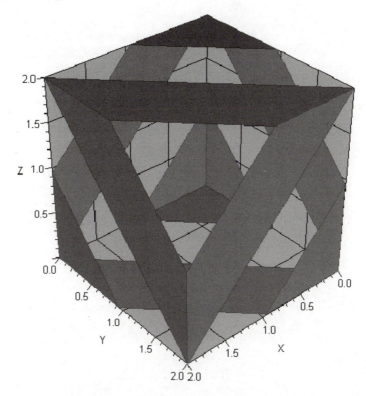

How many dimensions before we see (regular) octagons, pentagons, dodecagons, and so on? We won't! Even in an infinite-dimensional grid, the only regular polygons of finite size are triangles, hexagons, and squares. We could probably prescribe the nth coordinate of the kth vertex of a regular pentagon, say, but infinitely many coordinates would be nonzero, resulting in infinite size. And it would only *approach* regularity as we consider higher and higher dimensions.

Acknowledgment. Illustrations in this article were prepared with Macsyma.

Bibliography

[1] Marin Gardner. *The 2nd Scientific American Book of Mathematical Puzzles and Diversions.* New York: Simon and Schuster, 1961.

[2] Bill Gosper. "Rectangles Divided into (Mostly) Unequal Sqares." Available at http://www.tweedledum.com/rwg/squares.htm, 2007.

About the Authors

Vanni Bossi is the president of Milan's Club Arte Magica (CLAM). In addition to his magical performances and publications, he is engaged in spreading the history of magical culture.

Stewart Coffin is recognized as one of the world's best designers of polyhedral interlocking puzzles. He is the creator of AP-ART and author of several books on puzzle craft.

Frans de Vreugd is a Dutch puzzle designer and collector, traveling around the world in the search for puzzles. He is one of the editors of the puzzle newsletter *Cubism For Fun*.

Persi Diaconis is famous for combining mathematics and magic, with professional experience in both fields, and for studying the mathematical properties of such fundamental topics as coin flipping and card shuffling. With Ron Graham, he has recently written a book where mathematics meets magic, *From Magic to Mathematics—and Back* (to appear).

Jeremiah Farrell, an emeritus professor in mathematics at Butler University, is, with his wife Karen, editor and publisher of *Word Ways: The Journal of Recreational Linguistics* (a journal started in 1968 at the suggestion of Martin Gardner).

Martin Gardner is the father of recreational mathematics, most famous for his 25-year "Mathematical Games" column in *Scientific American*. He has written more than 65 books throughout science, mathematics, philosophy, literature, and conjuring.

Solomon W. Golomb, University Professor and Viterbi Professor of Communications at the University of Southern California, is an elected member of the National Academy of Sciences and the National Academy of Engineering. He has authored several articles in Martin Gardner's "Mathematical Games" column in *Scientific American*, and he is well known as the inventor of polyominoes.

Bill Gosper is a mathematician and programmer, recognized for his work in computer algebra, Lisp, and Macsyma. He is considered one of the founders of the hacker community, cowriting the notorious HAKMEM document. He also found the first infinitely growing pattern in Conway's Game of Life, the glider gun.

Ron Graham is famous for his profound contributions throughout combinatorics, coauthoring the book *Concrete Mathematics*, and using the largest number in a mathematical proof (Graham's number), as well as being an expert juggler. With Persi Diaconis, he has recently written a book where mathematics meets magic, *From Magic to Mathematics—and Back* (to appear).

Dick Hess is a designer and collector of puzzles and riddles, and a mathematician. He wrote a puzzle column, "Puzzles from Around the World," for ten years, and those puzzles are collected in his article in *The Mathemagician and Pied Puzzle*. He also has written the *Compendium of Over 10500 Wire Puzzles*.

David Klarner was an old friend of Martin Gardner, frequently contributing material to Gardner's "Mathematical Games" columns and books. Klarner edited a volume dedicated to Martin Gardner's 65th birthday, *The Mathematical Gardner* (retitled *Mathematical Recreations: A Collection in Honor of Martin Gardner* by Dover in 1998), and Gardner dedicated his book *Time Travel and Other Mathematical Bewilderments* to Klarner.

Ken Knowlton works in computer-assisted art and since about 1980 has concentrated on creating mosaics that integrate the material used and the nature of the subject. He received a PhD from MIT and worked from 1962 to 1982 at Bell Telephone Labs, during the formative years of computer graphics. In 1993, he created a domino portrait of Martin Gardner using six sets of double-9 dominoes.

Rodolfo Kurchan is a designer and collector of riddles and mechanical puzzles in Buenos Aires. He is the author of two books, *Diversiones con Números y Figuras* and *Mesmerizing Math Puzzles: Official American Mensa Puzzle Book.*

Mamikon Mnatsakanian is a physicist at the California Institute of Technology, who with Gwen Roberts has been fortunate to spend many pleasant hours with Martin, making tricks and exploring new possibilities for puzzles.

Christopher Morgan is a computer consultant, amateur magician, musician, and puzzle collector living in Boston. He is currently designing a website for book lovers.

Colm Mulcahy has taught at Spelman College in Atlanta, Georgia, since 1988, where he recently completed a three-year term as chair of the Department of Mathematics. He first read Martin Gardner as a teenager in Ireland in the mid 1970s but did not become besotted with mathematical card tricks until early 1999.

Prof. M. O'Snart lives in the attic of Toronto-magician Tom Ransom, an avid student of conjuring and its literature for 67 years.

Ed Pegg Jr, a former programmer for NORAD, is the webmaster of mathpuzzle.com and writes the weekly "Math Games" column for maa.org. He maintains library.wolfram.com for Wolfram Research.

Sir Roger Penrose is Emeritus Rouse Ball Professor of Mathematics at the University of Oxford. He is renowned for his work in mathematical physics and in geometry, in particular, for his famous tiles that force aperiodic tilings. He has written several books, of which *The Emporer's New Mind* received the British Science Book Prize.

Thane Plambeck is a technology investor and entrepreneur who lives in Palo Alto, California. He studied as an undergraduate at the University of Nebraska at Lincoln when David Klarner was a professor there, and he has a PhD in computer science from Stanford. When Klarner died in 1999, his wife Kara Lynn sent his unpublished works to Plambeck.

James Randi is a professional magician and author. The James Randi Educational Foundation is an educational resource on the paranormal, pseudoscientific, and supernatural.

Gwen Roberts is a high school teacher in Los Angeles. She creates and uses puzzles, games, and all types of hands-on activities to demonstrate mathematical concepts. Martin Gardner has been her inspiration and source.

Thomas Rodgers is the founder of the biennial Gathering for Gardner and a puzzle collector.

Wade Satterfield studied computer science at the University of Nebraska at Lincoln, where David Klarner was his PhD advisor.

David Singmaster is the principal historian of recreational mathematics. He was the leading expositor of the mathematics of the Rubik's Cube and a professor of mathematics at London South Bank University.

Jerry Slocum is the author of nine books about mechanical puzzles and is also known for his research on the history of puzzles and his large collection of puzzles and puzzle books. He is the founder and organizer of the annual International Puzzle Parties, held in the United States, Europe, and Asia.

Raymond Smullyan is a master of recreational mathematics and logic puzzles, having authored many books on these subjects. Many years ago, he supported himself as a magician when both he and Martin Gardner were fellow students at the University of Chicago.

M. Oskar van Deventer is the creator of hundreds of innovative mechanical puzzle designs, several of which are commercially available.

Rik van Grol is a collector, solver, analyzer, and designer of mechanical puzzles. He is the managing editor of *Cubism For Fun*, the English newsletter of the *Nederlandse Kubus Club*, an international club of puzzle enthusiasts founded in 1981.

Peter Winkler is a professor of mathematics and computer science, and Albert Bradley Third Century Professor in the Sciences, at Dartmouth College. He is also the author of two collections of mathematical puzzles, *Mathematical Puzzles: A Connoisseur's Collection* and *Mathematical Mind-Benders*.

David Wolfe is a professor at Gustavus Adolphus College. He coedited *Puzzlers' Tribute: A Feast for the Mind* with Tom Rodgers and coauthored *Mathematical Go: Chilling Gets the Last Point* with Elwyn Berlekamp and *Lessons in Play: An Introduction to Combinatorial Game Theory* with Michael Albert and Richard Nowakowski.